How to Build a Universe

A study of biblical creation
By David Glaze

Copyright © 2016 David Glaze

All rights reserved.

ISBN:1523419334
ISBN-13: 9781523419333

All scripture is quoted from the
King James Version which is in the
Public domain
Definitions of Hebrew words from
Strong's Dictionary which is in the public domain

Dedication

To all who hold a literal and historical viewpoint of the Holy Scriptures and particularly to those who believe in and defend the position that Genesis chapter one is a historical account of the actual creation event as revealed to mankind by divine revelation.

Epigraph

"The prophet that hath a dream, let him tell a dream; and he that hath my word, let him speak my word faithfully, what is the chaff to the wheat saith the Lord." Jeremiah 23:28

I Believe the Bible

I believe the Bible, every jot and every tittle,
I believe it from front to back and even in the middle.
I believe the Bible, that its sixty-six books were penned by the prophets of God and are worthy of any Nook.
I Believe the Bible even though it makes me tremble,
Its fearful words can be read in black and white or even on a Kindle.
I believe the Bible is the word of God and not of man,
It tells us of his will and salvation's wonderful plan.
I believe the Bible, I believe in its poetry and prose,
It is more correct that anything mankind could ever compose.
I believe the Bible more than what any critic has to say,
I believe it where it says Jesus is the Truth, the Life, and the Way.
I believe the Bible though heaven and earth may pass,
Its words are indestructible; they will last, and last, and last.
I believe the Bible it's a light to show the way,
It shines brighter than even the sun on a fair summer's day.
I believe the Bible even in our modern age,
Where computers and video games are all the rage.
As time goes on, I believe more and more – every sacred page.
I believe the Bible in spite of the spiritual darkness that abounds.
I focus upon its every word and pay close attention to all its sounds.
I believe the Bible, and in this, I am confident,
I know that every letter and syllable have all been heaven sent.
I believe the Bible, though some would call me a fool.
God has promised that if I am faithful to him, that someday over a portion of the earth I will rule.
I believe the Bible even though it says there is a hell for the lost.
If any man will believe the Bible he will escape this terrible cost.
Do you believe the Bible? Friend, I am not trying to put you down,
But if you will believe the Bible, then some day with me you can wear a robe and a crown.
I believe the Bible

Outline   pg#12

The Author of creation vs1
The beginning condition vs2
1. In the beginning
The moment when time and the universe began
The point of departure from eternity
2. God
Transcendent
Knowable
Personal
Mighty
The First Cause
3. Created
The modus operandi of God
The cause of all other causes.
4. The heaven and the earth
 Heaven – everything out there
Earth – everything here - {the central place}
The Beginning Condition Verse 2   pg#30
1. The original condition of the earth.
" *and the earth was*"
- Covered in darkness
- Formless
- Empty
- Water covered
- The presence of the spirit.
2. The methodology of the Lord.
- The key to the outline of chapter one.
God is going to:
- Light the earth
- Form the earth
- Fill the earth

Stage 1   pg#42
Day 1- Light Genesis 1:3-5
Creation on light
Observation and evaluation of light
Division of light from darkness
Naming of Night and Day
Noting the first day
The meaning of night and day
The basic cycle
The two modes
The 24-hour or solar day

Day 2 – Atmosphere Genesis 1:6-8   pg#53
Introduction
Study of the word "firmament."
The water canopy
The absence of the waters
The naming of the realm
The order of events

Stages 2 completed &3   pg#62
Day 3 Land and Sea and vegetation Genesis 1:9-13
Formation of Land and Sea
Naming Earth and Seas [a stage]
Evaluation of Earth and Sea
Creation of Vegetation [a stage]
Replication and multiplication of vegetation
Evaluation of vegetation & summery of first three days

Stage 4   pg#74
Day 4 Genesis 1:14-19
A change of perspective
 Why the sun, moon and stars were made the fourth day?
Why light the first three days?
The cyclic aspect of heavenly bodies

The light source aspects of the heavenly bodies

Stage 5   pg#88
Day 5 Genesis 1:20-23
Water and air creatures
Life? What is it?
Air creatures – Birds
The divisions of sea and air creatures
The word "kind"

Stages 6and 7   pg#96
Day 6 Genesis 1: 24-31
The creation on land creatures
The creation of man
Man – a creature patterned after God
The name of Man
The dominion of Man
Man and his gender.
Blessing and instructions given to man
The final evaluation

Day 7   pg#110
Summery and Sabbath day
A completed place
The seventh day

How to Build Universe Extras   pg#114
Keys of Genesis 1
A study of numbers in Genesis 1
The age of the earth
The" Rate" issue
The cancer Illustration

Exist

Is it at all possible for God not to exist?
Could there be no glory shore or land of eternal bliss?
Don't know about you but evolution seems much more of a myth.
As I look around one thought is quite profound.
The evidence of God does everywhere abound.
It stands there in each tree and lays there in the grass.
I see it in the wings of the bird that just flew past.
Hear him in the cricket which plays its song every night.
Or listen to the thunder as the clouds release there built up might.
The howl of the wolf, the shriek of a bat,
the song of the Whippoorwill asking his mate where are you at.
Each and every animal makes its own appropriate sound.
Doesn't that speak to you that the Almighty is in town?
Or what of the intricacies of the hearing ear?
Converting the airs vibrations into impulses by which we hear.
Each and every creature bears a mark of the divine hand.
This cannot be an accident it must have been a plan.
Consider a moment the inconspicuous tiny seed.
Who would have thought it bears within DNAs complicated creed.
What a marvel the seeds multiplying act.
And each and every one a copy that's perfectly exact.
Darwin was proved to have inheritance wrong but Mendel was on track.
The chromosomes he discovered are a marvel to unpack.
Or consider the atom, universal building block.
Its inexhaustible combinations keep the chemist on the clock.
Then there is the electron or the information stored in light.
Can you still think it's an accident not the workings of His might?

The truth is we observe a miracle world.
It's the only satisfying answer for the stars in space that twirl.
We see it in the moon when we watch it slowly rise.
And on the waves that push ashore upon the evening tide.
There is much too much order not to be a heavenly border.
And much too much design not to be one who created time.
And much too much coincidence for it not to be providence.
And much too much perfection for there not to be a heaven.
Oh atheist please pause and contemplate the grand First Cause.
Oh agnostic please take a moment recognize you need atonement.
The nature we adore is evidence of a glory shore.
The marvels we behold speak of a land with street of gold.
I wonder what would it take to end this great debate?
I fear that for to many the truth may come too late.
Many would say, "Well I must see God for myself"
But if you wait that long then for you there will be no help.
We all believe in things which we have never seen.
What wrong with believing this is all Gods universal scheme?
Go ahead risk it, dare to believe.
God has a gift for you just reach out and receive.

The Author of creation vs1

"In the beginning..."

This phrase denotes the furthest back point, which can be made reference to. Before this is eternity past. Before this is the spiritual world of God and the angels. Before this there was no material universe. Not a speck of dust, not a cloud, not the minutest particle. A nothing of nothing. No singularity.
Note the definition of the word as it is used in Gen 1:1
re'shit [ray-sheeth] 7225, "beginning; first; choicest." The abstract word re'shit Connotes the "beginning" of a fixed
...." This word can represent a point of departure, as it does in Gen. 1:1 (the first occurrence)
This is the moment that there was the first something . This the moment that there was a departure from the eternal past. Beginning means beginning just as we would use it in everyday language. As the beginning of a day, or week or year or anything else.
In this phrase and in this verse the answer to one of our deepest question is answered. This where it all came from. This is how it began. The universe begs the question of where did it come from. Because it exists and because we exist, we ask the question. Specifically the phrase " in the beginning" tells us the when.. Time itself can be understood to start here at this moment. With the first act of creation corresponds the first tick of the clock. There is nothing to record before this. Time did not exist before this. Time as a thing is connected to the universe. Time is a created thing. Time is a property of the universe. A special property and therefore before there was any space or any reference points there was no time. In addition, this tell us that eternity is a timeless state There was no time in the since of the passing of time before Genesis 1:1 There was nothing for it to pass from or to. God has existed in a state beyond time.

This phrase "In the beginning" tell us something about "time", about "eternity" and about God. This beginning point is also a final point in our search backwards . The end of our quest as it were. No further questions should be asked about the " when". We should find a since of satisfaction and of finality. We should come to a sense of mental rest and of knowledge. This was the beginning.

Also this phrase controls the context and the back drop verse one .The context of verse one is crucial to what we learn about God from this verse. Moreover, the stage of the whole chapter is set by this phrase. The whole of chapter is to be taken in reference to this.

Note how 1:1 related to chapter 2:1

'" In the beginning God created the heaven and the earth."
and *" Thus the heavens and the earth were finished, and all the host of them."*

What the writer begins to explain in 1:1 continues to 2:1

The subject and the context are the same. There is no break in between. To propose a break between verse one and two is just bad exegesis. Also to say that this is not the original beginning of all beginnings of the universe is also to interject something into the text and to denies the meaning of the words that are here. Another thing is that if this is not the original beginning then we are left in the dark as to the original beginning.

" God"

This section is short study of some of the attributes of God as they relate to creation. God is the essential ingredient in creation. Creation is a reflection of him. It is his being and nature which provide the knowledge, power and skills which were required to produce such a grand edifice as the universe which we live in. God is the architect. God is the engineer. God is the financer. God is the laborer. God is the dreamer which conceived and concocted it. God's hands molded and shaped it. From the drawing board to the last finishing touch it is his conception and his alone. He asked no help. He needed no help. Of all the things we say he is "Creator," is the First. Creator is the first point on his resume. It is fitting that a review of his attributes be included here at his first mention. Creation could not exist without the Creator. Surely the events recorded in Genesis chapter one could not have happened without a God who is everything and more that what is written here. But also if God is everything that is written here then the events of Genesis one can be conceived of as an orderly and masterly executed plan.

God is transcendent.

That is outside of the universe. Not a part of the universe. This is implied by the context. Because the context is the beginning of the material universe and we see that God already exists then he must have already existed outside of the universe. The context is very important here. The context sets the scene. The context draws a sharp line between what was about to happen and what was already there.

— tran·scen¹dent·ly advertency noun

"Being above and independent of the material universe. Used of the Deity."

[The American Heritage Dictionary of the English Language, Third Edition is licensed from Houghton Mifflin Company. Copyright © 1992 by Houghton Mifflin Company. All rights reserved].

. That God exists without the universe and outside. Nowhere is this idea shown better that in Genesis 1:1. He is totally other than we are in his being. He is without a need of the universe. Without the need of a habitat. Totally other and totally self-sustaining and immaterial. This is why the scientist cannot directly find him or observe him. The universe is his creation. He is not in it or a part of it. No more that the inventor in part of his invention. God exists in a realm beyond our examination.

God is Knowable
But this is not to say or mean that we cannot know of him or understand him. There are three ways that we can know of him.
1. By his actions.
2. In history
3. Personally.

We know of God by his actions. Because he has acted in the creation of the universe and in the creation of ourselves. God has not just done something in the universe but he has done ever thing in the universe and we can know him by these actions. His power, knowledge, ability, engineering skills. are known by his action of creating. Psalm 19::1 *The heavens declare the glory of God and the firmament showeth his handy work*. The scriptures confirm the argument of the " watchmaker". That is if it is true that there is a watch then there must have been a watchmaker. The work speaks of him and if we attribute the work to some other then we are in denial of him. See Rom 1:19 -20 God himself expect that we can recognize this and it is as reasonable to expect that God can be known by his works, as it is when we attribute any other act to the creature or source that caused it. Be it a person or animal of force. It truly is no secret that God has acted we see it everywhere we look. Whenever we speak of nature or Mother Nature, we are speaking of the works of God. This is an application of the principle of cause and effect, which states that for every effect there must be a cause.

Secondly, we can know him through history. This is chiefly found in the history of the Jewish people and the prophecies of the Bible. The prophecies that have been told and fulfilled. Many of these can be proved historically. Such as Daniel's forth telling of the kingdoms between his time and Christ's time in chapter two. Also in the prophecies of the first advent of Christ which were fulfilled in his life, death and resurrection. The only way to deny these things is to deny history. A person must deny nature and history to reject the truth of God.

Thirdly, we can know him by personal experience. This is done through the new birth of John chapter three. Each person who comes to accept him a savior does experience knowledge of God for himself or herself. God makes himself known personally to them through the Holy Spirit. This is a real experience. Just as real and just as valid as the first two. God can and does interact with us personally just as he can act in creation and just as he has interacted with the nation of Israel in history. God can be observed in our lives and in the lives of others.

God is Personal.

God is the great, great, spirit person who is out there.

This answers the question of "whom". This is the point of departure as far as the naturalist is concerned. Because the naturalist rejects anything that has to do with a personal God. The word God is a personal pronoun. Therefore, his personhood is implied thereby. Any other interpretation is a denial of the language. Science presuppositions that this cannot be or that it is better for human advancement to never look at things this way. Einstein admits that the possibility of God cannot be ruled out. I quote :

" To be sure the doctrine of a personal God interfering with natural events could never be refuted, in the real sense, by science, for this doctrine can always take a refuge in those domains in which scientific knowledge has not been able to take foot"

OUT OF MY LATER YEARS page 26

God as a person cannot then be disproved. Yet Einstein rejected this idea. Again, I quote:

"In their struggle for the ethical good, teachers of religion must have the stature to give up the doctrine of a personal God, that is, give up the source of fear and hope which in the past has placed such vast powers in the hand of the priests."

For human good then would be the reason that he rejects God as being personal. However, this is not a scientific reason. Only a practical reason based on what Einstein view of what religion has done for humankind in the past.
What is our greatest argument that there is a personal God?
God has spoken – inanimate forces do not speak, think, or show any of the attributes of a person. Along with the idea that God has spoken I would bring in the idea that Genesis is not a theory or a conceptualization of man. But is divinely revealed facts. These are statements of facts and of actual history. To this, again Einstein would disagree. Because he believed that religion *"deals only with evaluations of human thought and action: it cannot justifiably speak of facts and relationships between facts. "and that " a religious community[should not] insist on the absolute truthfulness of all statements recorded in the Bible"*
To this, I totally disagree. To this, I propose the opposite. We believers must hold the position that the bible is accurate and correct in all points of history and science. And in all points period. We must hold this ground. All else will ring hollow. There is no other position. The believer stands or falls upon this doctrine.
Jer 23:28 *The prophet that hath a dream, { or scientist } let him tell a dream; and he that hath my word, let him speak my word faithfully. What is the chaff to the wheat? Saith the LORD.* (KJV) Jer 23:29 *Is not my word like as a fire? saith the LORD; and like a hammer that breaketh the rock in pieces?* (KJV) The scripture claims superiority. If it is what it is then it is superior or it is nothing and we are fools. No theory shall stand that contradicts the word of God.
We must not take lightly this idea that God has spoken. God has left us a message this is how we know he is personal. His message bears the mark if its author just as his work bears the mark of its author. God has communicated to us through the same means that we communicate to each other through words. And he expects us to accept his message. Another scripture make this point clear" Luke 16:29 *And Abraham said*

unto him'" they have Moses and the Prophets let him here them," scripture is greater than nature. Scripture tells us what nature and forces can never tell us. Scripture is more direct that nature. You cannot get any more direct than words. If we deny his words, how can we hope to find him elsewhere? Especially since, he is outside the universe. We can discover what he has done and then we can attribute it to happenstance or we can attribute these things to him. Yet until you will accept his words, you cannot come to know that he is personal. God has written to us this is how we know he is a person. The scriptures bear divine evidence. An honest study of the scriptures will demonstrate this divine evidence. Genesis 1 is divine evidence. No human writer could have known the things told here. Many of the things told here were unknown until modern times. Yet this is an ancient book. Certainly, no one would date Genesis in modern times. Therefore, how can you explain the things revealed here except by revelation? All of a sudden, this becomes the grandest issue of all. The most crucial, the hinge upon which everything else swings. The water shed of all watersheds. The point of departure or the point of acceptance. God is a person and God has spoken. God has given his opinion God has stated the facts on certain things. And the record of this speaking is what we have in the Bible. And we are able to understand these words. In direct and simple language; in common human language. Genesis 1 is not the language of the scientist or of the intellectual, or of the poet but is common everyday language that all who can read on a normal level can understand. He has not spoken to us over our heads or under our feet but upon our level.

God is mighty.
. His name means power. A plurality of powers.
Strong's # 430 'elohiym (el-o-heem');
Plural of 433; gods in the ordinary sense; but specifically
used (in the plural thus, especially with the article) of the
supreme God; mighty.

There is another passage, which explains God to us the power of God in the simple and most profound way Exodus 3:2
" *behold, the bush burned with fire, and the bush was not consumed.*"
This is the picture of his power. His power is intrinsic, self-sustaining, Eternal everlasting power. The scientist should find an interest in this verse because it has standing in antithesis to it the knowledge of the way that energy works in the material universe. In the same way, that Moses is amazed and interested in how it is that this bush did not burn so the scientist should be in God and his power. The scientist should also find a satisfaction in this verse. This is an adequate answer for the scientist. A principle of energy that is unknown to him but is reasonable to explain all the rest. If God has intrinsic energy and $E=mc^2$ then this is the perfect solution for where the universe came from. God burns but he is not consumed. He expels energy but it takes nothing away from him. His energy is never less after he has performed some act than it was before he began. He is inextinguishable. Just as is his eternality so is his strength. It is a continuous unchangeable aspect of his being. Because of the strength of God, he is an adequate first cause.

Isaiah 40:21 *Have ye not known? have ye not heard? hath it not been told you from the beginning? have ye not understood from the foundations of the earth? :22 It is he that sitteth upon the circle of the earth, and thinhabitants thereof are as grasshoppers; that stretcheth out the heavens as a curtain, and spreadeth them out as a tent to dwell in: 40:28 Hast thou not known? hast thou not heard, that the everlasting God, the LORD, the Creator of the ends of the earth, fainteth not, neither is weary? there is no searching of his understanding.* (KJV)

First cause.
God is called the first, Isa_41:27 *The first shall say to Zion, Behold, behold them: and I will give to Jerusalem one that bringeth good tidings.* Isa_44:6 *Thus saith the LORD the King of Israel, and his redeemer the LORD of hosts; I am the first, and I am the last;*

and beside me there is no God. And one more, Isa_41:4 *Who hath wrought and done it, calling the generations from the beginning? I the LORD, the first, and with the last; I am he.*

Four principles about the first cause
1. Something has to be eternal
2. This eternal thing has to be adequate
3. The first cause is unexplainable
[To be able to explain the first proposition would be a contradiction]
4. There is no further step backward

Genesis 1:1 implies that God is eternal. Before the beginning, he was already there. This goes along with his transcendence. If God was there before the beginning so that he might act in the beginning then this shows that he is eternal. Eternality is the state in which God exists. A state of timelessness. With God, time does not go back or forward. He is on the outside of time just as he is on the outside of the universe. He sits as an observer of time. Time is one of his created things. God is this something that has always existed. He is the answer to the question of what has always existed, of what is there that is eternal. The logic that leads us to ask this question finds its answer in him.

Isa 48:12 *Hearken unto me, O Jacob and Israel, my called; I am he; I am the first, I also am the last. 13 Mine hand also hath laid the foundation of the earth, and my right hand hath spanned the heavens: when I call unto them, they stand up together.* (KJV)

The universe begs the question of where did it come from. What Something had to be first. Something has to be eternal. Why? Because it is not logical that the universe could come from nothing. It is not possible, and therefore not logical, that something can come from nothing. Right? Now God proposes to create something from nothing but this is not a nothingness of nothingness because you have God and his power present to do this.

The question can be asked why is there something instead of nothing. There are three possibilities.
1. Either everything came from nothing.
2. Or everything came from something.
3. Or everything is eternal.
If you rule out that first answer on the basis of reason,[something cannot come from nothing] and you rule out the third answer based on history and on the second law of thermodynamics, that is that the universe is running down. Then what is that something that was first? The scriptural answer is the eternal God person..
The text that goes along with this is Exodus 3:2-3 and 14. The "I AM THAT I AM"
The bush that burns but is not consumed this is the first cause. The cause without a cause, the eternal thing, the first if all firsts. The root, the spring, the source, the prime mover. The question of where did it all come from stops here. This is the adequate explanation. This is the unexplainable. God does not even explain himself HE just is that is. This is the first and final answer. This is the only answer. This answer is most basic and most daunting.
In addition, this is an adequate cause. It is not enough to just say that something is eternal. This something must also be adequate Something that is able to account for the rest. To say that there was just matter or there was just energy is not an adequate answer. Energy and or matter do not explain the rest. Such as life or order or design. But the eternal God person is an adequate answer. All else can be accounted for through him.

Now that he remains unexplainable is actually logical. Because there has to be something that is, first and therefore something that is unexplainable. To explain where This first thing came from would be to contradict the principle that something had to be first. In addition, if you cannot say that something was first then you would be stuck in a loop of forever asking the question of what caused that? and them again what caused that? on and on forever.

The last issue on this point is that this is the final step backward. This is the resting point. This is a satisfactory answer. This is a final answer. It is not reasonable ask the question of what was there before there was the eternal God person. This goes against reason and against logic. God is eternal and that is as far back as you can go. Eternal is eternal. We need not ask further because there is no further to be asked about. The answer needs to simple be accepted. The mind that longs for the answer to the existence of things is answered in him.

"Created"
The modus operandi of God
The cause of all other causes

Only God can create in the sense that the word is used here. How do we understand this aspect of him? The fact that the Lord uses no mechanisms and no processes and that he needs nothing to do anything. How do we explain the command - done relationship? The voice power or the will power of God? This is as inexplicable as is his eternality. This also is so "other" than we are. Now we know that his power is inexhaustible [ex 3] but how does his power work to bring things into existence? As far as I know, there is no passage that explains this or gives us any clue as to how God does this. Jesus said he used the "the finger of God " Luke 11:20 to do his miracles. Then what does this not explaining tell us? This tells us that this ability to create is one of the unique things about God. It is to be seen just as his being. God has a total different ness of doing things than his creatures. This ability to create is to be understood just as his being, eternality and power are to be understood. God is the, "I AM THAT I AM" and also he is the I CAN AS I CAN. The mystery of this is to remain forever. This God keeps reserved for himself. This is as much a part of him as is his being. God is the inexplicable person who has the inexplicable ability to create. This is who he is. This is part of what defines him. Creator is his middle name as it were. There is no more explanation for it than there is for his existence. We should admit and allow that some things are beyond us. And that this is okay. This is right. Let is accept our place as creature. Let us accept God's place as creator and everything else will be all right. The Lord has the ability; he has a way to call things out of nothing. Maybe it is a transformation of some of his energy into matter and then to shape and form it according to his knowledge and wisdom. This might be the best answer that we can give to the scientist who is looking for the means and the mechanisms of nature. But his energy is still not the same type of energy that we know of. His energy is of some other type. It is a spiritual type of energy and a type of energy of his being. Hebrews 11:3 may be a help to us along these lines.

"Through faith we understand that the worlds were framed by the word of God, so that the things which are seen were not made of things which do appear."

This says to us pretty much the same thing as Genesis 1:1 does. The difference being Genesis 1:1 is a statement of fact and Hebrews 11:3 being a statement of how we are to accept this fact. Through faith that is. The bottom line is still a non-natural cause or to state it positively a supra natural cause. The "natural" cause, which the scientist is on a quest to find, does not exist in the area of origins.. [This is an assumption based on faith] To prove origination as the scientist seeks to do and to prove origination as the thing that the scientist would accept is impossible. You cannot prove a natural means if there is no natural means to prove. And you cannot prove a supernatural means because there is nothing left behind to examine. In other words, there is no natural law or force involved that the scientist might examine through scientific method. Demonstration is not possible. It is not within human power. It does not lie within the natural realm, which is the only realm, which we can examine.

Faith is the only answer. Faith is the only means of understanding and of acceptance of the proposal of a something from nothing situation. A math illustration here might be helpful. Zero plus zero is zero. That is the problem with the something from nothing situation. How is it that we can get beyond the sum of zero?. But zero plus "bara" equals something. This is how we get past the sum of zero. This "bara" being the X factor of the power of God. "bara" cannot be directly observed. So we come back to the proposition that we must accept "bara" by faith. This is a requirement that God places upon us. Hebrews 11"6 speaks of this"

But without faith it is impossible to please him, for he that cometh to god must believe that he is, and that he is a rewarder of them that diligently seek him".

Now it seems at first to be a weak position that a presuppositions such as faith lies at the bottom of all of this. However, I think we should view faith in this context as the opposite of naturalism. Naturalism presupposes that we live in a closed system. That is that nature is alone. This is also a presupposition, an assumption on the part of the naturalist. Faith is at least an equal proposition to naturalism or to atheism. Faith proposes that we live in an open system. That God, the transcendent one, crosses the system. That the system has been created by the outside. This is what God has done by "bara" and by his interaction with man in history. This is just as legitimate a position as naturalism. Faith is not a foolish proposition. Faith is just one of the two only choices that there are. We may choose to believe in a closed system or in an open system but both is a matter of presupposition.

The people who started the modern scientific method did so because they proposed that God acted according to reason and that therefore the universe could be understood through reason. Then somehow, the scientist and philosophers who came along later concluded that the proposition of God was not a reasonable idea. Therefore, the modern scientist has the presupposition of a closed system. Modern theologians also presuppose a closed system. Genesis 1:1 shows that there two groups are wrong. The first verse shows that we live are an open system between that which is outside, God that is and that which is natural, the universe that is. To hold to the position of a closed system is a violation of scripture.

This most important of all presupposition is a heart matter. A matter of how the heart chooses to interact with the world or how the heart decides to place itself under the restraint of the Deity or not. Actually, this choice is a privilege of our autonomy that we have because we are made in the image of God. God has sovereignty and autonomy and we have autonomy or self will also. Otherwise, we would not be in his image. It is ironic that one of the very things that are our greatest privilege is also, what will lead many to damnation but this is the privilege and the responsibility of being autonomous.

Now to return to the word create and look at its definition and usage. CREATE, TO

Bara' 1254, "to create, make." This verb is of profound theological significance, since it has only God as its subject. Only God can "create" in the sense implied by bara'. The verb expresses creation out of nothing, an idea seen clearly in passages having to do with creation on a cosmic scale: *"In the beginning God created the heaven and the earth"* Gen. 1:1; cf. Gen. 2:3; Isa. 40:26; 42:5. All other verbs for "creating" allow a much broader range of meaning; they have both divine and human subjects, and are used in contexts where bringing something or someone into existence is not the issue. Bara' is frequently found in parallel to these other verbs, such as `asah, "to make" Isa. 41:20; 43:7; 45:7,12; Amos 4:13, yatsar, "to form" Isa. 43:1,7; 45:7; Amos 4:1, and kun, "to establish." A verse that illustrates all of these words together is Isa. 45:18: *"For thus saith the Lord that created* [bara'] *the heavens; God himself that formed* [yatsar] *the earth and made* [`asah] *it; he hath established* [kun] *it, he created* [bara'] *it not in vain, he formed* [yatar] *it to be inhabited: I am the Lord*; *and there is none else* A careful study of the passages where bara' occurs shows that in the few non poetic uses (primarily in Genesis), the writer uses scientifically precise language to demonstrate that God brought the object or concept into being from previously nonexistent material.

Though a precisely correct technical term to suggest cosmic, material creation from nothing, bara' is a rich theological vehicle for communicating the sovereign power of God, who originates and regulates all things to His glory.

The claim that God created the universe runs throughout the bible from beginning to end. If this claim cannot be trusted then none of the bible can be trusted. God created, this is the explanation for how universe came into being.

Nature is a great schoolhouse for us and it is noble and proper to study and learn from it. The scripture says this. Psalms 19:2 *"Day unto day uttereth speech and night unto night showeth knowledge."* Let us learn all that we can from nature. That is what it is there. Nature was not only made to sustain us but also to educate us. Science would be far better served if the aim of origins were set aside and nature was studies for what it is now and for what there is for us to learn from it and do with it. This is how we might best benefit ourselves from nature. The origins of things have been divinely declared unto us. We need not make further inquiry as to their origination. God has also caused all other causes. What does this mean? This means for one that there are no accidents. Nothing that exists is happenstance. All forces, powers, all bodies, all arraignments and locations of things. All orders of creatures. All inter relationships among things. All that we find in nature is the result of the direct command of God. The things of nature are of divine ordination. All the forces of the natural system - the force of gravity, electricity the strong and weak nuclear forces. The constants such as the speed of light etc. These are of his design. The way the universe works. Its balances and intricacies. All these things are here for our benefit and for our admiration.

"The Heaven and the earth" This phrase tells us what it is that God created. Note the division. Two distinctive areas or things. Heaven and earth. A more general term might have been used. It could have used the words 'all things" or "universe" or the word "everything" or some such idea.

Especially since this is a general and overall type of statement but it does not. This is for a reason. The reason being the significance of the earth and the centrality of the earth in the plan of God. Heaven is separated from the earth because one ,it is a different type of sphere. That is a gaseous or empty space as opposed to the solid earth and secondly in terms of importance. The word heaven in this verse speaks mainly of what we call outer space, the region of the sun moon and stars. This in and of itself is importance because it denote that there is a difference between the two. Heaven and earth are not the same thing. This same word is also used of the atmosphere and of the heaven that is the place of God abode. The word earth here is just as importance. The earth as compared to the rest of the solar system and the rest of the universe seems to be a small and insignificant place but in terms of its place in the plan of God this is not so. The earth may not be at the center of the solar system as well known. { The scripture never said that it was.}The earth is central or the center as far as God's plans and activities are concerned. This centrality of the earth is the main point to be noted. Note how this related to the rest of the chapter. The details, which we are given, are mostly about the earth. This is not because the author of Genesis did not know the comparative size of the earth to everything else but it is because of the importance of the earth. The stars are mentioned in just one phrase and the sun and moon in just a couple of verses. But the whole is concerning how everything relates to the earth. The setting up of the earth system that is the home of man. The earth is the center stage of the universe. The earth is the place where all the happening takes place. The earth could be compared to your favorite "hot spot" to the place where the action is. Here on our tiny island our tiny oasis within the great expanse of everything else. It is here that the grand issues are carried out. It is even here where God shall make his eternal abode when the new Jerusalem shall come down out of heaven. Here the drama of the fall and of redemption takes place. The earth is the central place.

The beginning condition vs2
And the earth was without form, and void; and darkness was upon the face of the deep. And the Spirit of God moved upon the face of the waters.

The original condition of the earth.
" and the earth was"
- Covered in darkness
- Formless
- Empty
- water covered
- The presence of the spirit

2. The methodology of the Lord
- The key to the outline of chapter one

God is going to:
- Light the earth
- Form the earth
- Fill the earth

The snapshot Genesis 1:2

God chose this as the starting point. The earth is the starting point. The place where he would begin the account. The earth also stands at this time alone, by itself in creation. There is nothing else in existence as of yet. Only the raw material and the foundations are in place.{ Job 38 :4-7} Note that the original snap shot is taken from outside. As if some third party observer were present on the scene and a witness to the first act of God. Now of course God is watching himself as it were. He is the recorder of his own first actions. How this first observation is made is interesting. Because all is in darkness the earth could not be seen, nor the Spirit, nor the waters. This first observation is of divine origin. Only God was there and only he could have known. Only he has the ability to know these things. Psalms 139:12 " *Yea, the darkness hideth not from thee; but the night shineth as the day : the darkness and the light are both alike to thee."* All the information from 1:1 to after the creation of man is totally of divine origin. God records what only he could know. Bring on all the other religions with their stories of creation and bring on all scientific theories and let us compare. There is no story like this story. There is no revelation like Bible revelation. This word, this message shows itself to be of divine origination.

How can we grasp the beauty of the first moment? The things that were there and the thing that were not. Just God about to something. God is on the move. God is in action. Surely it is a moment like none before and none after. When nothing moved but the Spirit and nothing observed but the Father and nothing acted but the Son. {See John 1] What a peek at God being God. How unique and wonderful are the scriptures! There is no way to study backwards and come to the conclusion of 1:2 as the first moment and the first thing that existed in the universe. The earth has been changed to many times since then. First by the acts of the rest of creation week. Then by the flood of Noah's time, and by the braking apart of the first continent. Finally by the daily processes of wind, rain and erosion. The cloud between then and now is far too dense and varied for man on his own to have ever come to this conclusion. Just look as the thing which man proposes. Man does not even start with the earth but first with a singularity and then some dust cloud, which condenses into the solar system. Then consider how the earth was a spinoff of the sun and on and on as the naturalists imagines it. Then once you had the sphere of the earth it would have to develop into the conditions for life. But God has given us as actual record, an actual account. An actual description of the earth in its first moment.

There are three main conditions noted of the primordial earth.

It first noted that the earth was "without form". Secondly it was "and void "thirdly there was "darkness upon the face of the deep" .And finally there is noted the presence of the Spirit. These three things then are the key to the arraignment of the rest of the chapter. These three things are what God is going to deal with. These three things reveal the outline of the rest of the chapter and the rest of creation week. God would deal with the earth being unformed, empty and dark. Each of their conditions would be addressed. Each of these conditions would be dealt with. Each of these conditions would be undone. Each of these conditions would be reversed. Whereas the earth was unformed it would be formed and arranged. Whereas the earth was empty it would be filled. {This also includes the heavens} And whereas the earth was dark it would be lighted. This was the plan.

The methodology of God
The primordial earth was therefore created as a sum of the raw materials, which God would then mold and arrange into shape. Just as any builder must first gather together the raw materials of whatever type and amount for whatever project that he has in mind. God followed an orderly and logical process in creating the earth. God is methodological. He did what any body would do who set about to build something. The Lord followed a plan of making things in mass and then bringing them into refinement. He shaped and arranges after he has created the initial mass of things. He makes the pieces and then puts them in place. As if the universe was some gigantic jigsaw puzzle. He acted reasonably and logically. The difference between him and us is in degree and scale. He did what anyone would do who had the power and knowledge that he has. Also concerning his method of doing things it can be noted that each day has two major events accomplished.. Now there being six days of creation and two major events for each day makes for twelve major stages or twelve major events in creation week.

What an injustice we do to the scriptures by presupposing that God must all ways do everything that he does completely and instantly. That he would not act in an orderly way or that he would not act methodologically, according to a process or a series of stages. By interjecting this idea of incompletion or imperfection into verse two the whole of the plan is missed. Verse 2 is not about chaos but about a plan. The logical plan that God followed. He created the raw materials and then he formed them into what he wanted them to be. Just look as salvation for an example. God saves us according to a plan. Certainly we are not what we are going to be. This is an example of God not yet completing a work that he started. He will complete our salvation just as he would complete the earth in creation week.

God followed two overriding methods in going about his work.

First is the principle of forming and filling. This sets the chapter into two major section of days one through three and then four through six. This first principle is drawn from the study of the words "without form and void "in verse two.

The second is that of Arraigning drawn from the word "divide" and "divided" as used throughout chapter one.

This is how God went about setting up house. This dividing has to do with setting things in their proper places and proportions. For example he created the water and then separated it out into the portions. Some above, some upon and some below the surface of the earth.

The one main substance.
Water

Along with the things mentioned above is mentioned also that the earth was covered in water. Earth is the water planet. Water as the main substance. Water, which is so unique to the earth. Why was water the choice material? It could be because the earth is made up of water more than anything else. God would have to change water the lest as opposed to some other situation such as starting with a dry crust an then afterwards making the oceans. To God water made the most sense and was the most usable and practical place to start. The first reasons for the water planet then were because of the volume and because of the practicality of water.

Another reason for water being the prime substance has more of a spiritual reason behind it. This is the idea of birth. The earth was "born" in water because this is how we are "born". The earth was born in water just as we are born in water. This is the birth of the earth so it is befitting that water is present. The earth is God's baby!

The condition of the water
The temperature of the water is also an important aspect of the original condition.
The water here is in liquid form. It must have therefore been in the temperature range for the liquid state. Now you might think with the other condition such as the darkness and the absence of light that it would be near absolute zero but this is not so. The earth was then much warmer than the space around it. The original water was created in the liquid state and had enough warmth in them to remain that way until the light was created.

The water also had a considerable depth to them. The word "deep" is where this comes from. Now how deep is deep? We are not told but deep is to be opposed to shallow and concerning the oceans this could be anywhere from a few hundred feet to several thousand feet. In addition, this original depth was soon to be changed on day three anyway. We are given only a general idea but this is enough to describe what God wants us to know of the first conditions. We may also assume that this is salt water since this is the type of water the oceans have in them to day. There is no need to account for where all the salt came from as the naturalist must do. The waters they were warm, deep, and salty. Also they were still and empty.

Darkness
"Darkness was upon the face of the waters." Darkness meeting water was the only distinction that could be made in the Ge.1:2 situation. The darkness that existed then was somehow different from the darkness that we have now. You can see this in what happens on day one. The darkness here is the eternal nothingness that had existed for eternity. It is an absence of anything whatsoever. Even empty space has certain properties but not this darkness. This shows us what there would be without God. If God were not there to act then all that there could ever have been was this darkness. A darkness with no space or time, a nothing of nothingness. The philosopher has asked the question of why is there something instead of nothing. The answer to the question is because God exists. Because without God there is only this nothingness. This nothingness, which the philosophers logically proposes that there would have been. Without God there really is no reason for anything to exist. Francis Schaeffer in his book, The God who is there, speaks about the importance of antithesis in our thinking process. And one of the things he says about this is," the greatest antithesis of all is that he exists as opposed to his not existing; He is the God who is there. This is the truth of

what the darkness was all about. Without God only darkness. However, with God in the equation anything is possible.

This same thing is true of people and their inner life. Without God there is darkness both morally and in any real hope for the future. Death and meaningless are the only thing that man has to look forward to without the hope of a God who is there and who is benevolent. Both of these things are found in the God of the Bible. God is light and in him is no darkness at all. He stands as the great antithesis to the darkness. This is true in regards to what would actually exist without him and of humankind's outlook upon life.

Without Form

Strong's # 8414 tohuw (to'-hoo);

from an unused root meaning to lie waste; a desolation (of surface), i.e. desert; figuratively, a worthless thing;

adverbially, in vain: KJV-- confusion, empty place, without form, nothing, (thing of) nought, vain, vanity, waste, wilderness. (DIC)

In the context this word makes perfect sense. The globe is there but it is in the state of raw materials. Like having a stack of wood that noting has been made out of yet. The shape and form has not yet been made. There is no dry land, no air and the ocean basins are not in their final position or in their final arraignment as of yet. The earth was just as this was translated, without form. It originated in an unformed condition. This word has the earth as a habitat in mind. The plant life is included is this. The preparation of the habitat is in view. If we were to establish a zoo or consider having some animal we would first have to have a place to keep it. This place would have to be a suitable habitat for whatever creatures we intended to keep. We would have to take into consideration all the things that such creatures needed. Air, water, temperature, light, shelter, food, surroundings, plants, behavior, interrelationships with other creatures in short the entire ecosystem.. In dealing with the unformed earth God in would take in consideration every individual creature that he intended to make then prepare a place for each one.

As you progress through the chapter the word "divide" is mentioned verses 4,6,7,14 and 18. This speaks to God going about to arrange things, to set them in order and in place. This is one of the keys to the process the Lord would follow. This stands in opposition to the formlessness. God would divide:

The light from the darkness vs.4

The waters – taking some from below and placing them above. vs.6&7

The day from the night. vs.14&18

There are other aspect of this going about to arrange things that do mention this word but are the same in action. He caused the dry land to appear and placed the oceans in there basins.

All of this is part of the doing away with the formlessness. The Lord is the great architect and the landscaper , framer ,developer of heaven and earth. His hands have shaped the world we enjoy.

Void

Strong's # 922 bohuw (bo'-hoo);

from an unused root (meaning to be empty); a vacuity, i.e.

(superficially) an undistinguishable ruin:

KJV-- emptiness, void. (DIC)

Since the word chaos is often used to describe what the Bible means by void, it should be noted that only the original meaning of the Greek word chaos (vast space) applies. Our modern understanding of chaos as confusion is not the meaning intended by the biblical writers.

(from Nelson's Illustrated Bible Dictionary)

(Copyright (C) 1986, Thomas Nelson Publishers)

This is the final condition that is that is described to us in verse two and the last thing that God would deal with. Void here means empty and nothing else but empty. This fits the context. This is the condition of the planet in verse two the inhabitants are what are in view at this point. There were no animals, no living creatures. After creating the habitat God would fill the place with creatures. Each realm, air, water, land would receive their intended creatures. There is no implication of any other activity implied by this word. The world was created empty because the world was not yet ready for creatures. The preparations had not yet been completed. The earth is the only planet that we know of that has been completed. The earth is the only planet with any living things. Science has discovered many things that have change our outlook upon the privileged position of the earth in the cosmos. Such as discovering that the earth in not at the center of the solar system, or that there are other planets. Yet when it comes to life forms the earth still stands unique in the universe. Try as we may to unravel the mystery of life we have not done it. Moreover, try as we may to find it somewhere else we have not done it. Science is on quest to do this. Naturalism is at a dead end on this point. Now if life were found somewhere else it would not disrupt the biblical scheme of things. The scriptures are silent concerning life elsewhere in the universe. If you did find it, it wound only mean that God had placed creatures somewhere else and had not told us about it. My opinion is that he has not. I believe that the biblical scheme of things is for life on this planet only. At least in the present age. Romans 8: 19-22 says that the whole creation is effected by fall of man. The whole of creation is waiting for the redemption of man. This would also apply to life outside of our planet or it implies that this world is the only place that has life forms. The other planets in our solar system are not fit for life and everything else is too far away to verify. The things that were said about the darkness also apply to the emptiness. That is if it were not for God's intervention that is the way it would have remained. God filled the earth with its creatures. That is how the scriptures says they came into being. This is opposite of what evolution claims. It is a contradiction in terms to say that God could have used evolution. The two ideas are polar opposite. Both cannot be right. Let the battle lines be clearly drawn. Darwin himself knew what his theory implied concerning its religious implications. He knew that what he proposed was a totally naturalistic means of the origination of life forms. This is why so many joined the cause with him so quickly.

And so was born our whole modern secular evolutionary society. But let the bibles position and the believer's position be clear. The earth was empty and God filled it.

The Spirit

"...and the Spirit of God moved upon the face of the waters."

The last thing that is noted of the first moment is presence of the Spirit poised for action. God in his essence is a spirit being. John 4:24 "God is a spirit ..." The actions of God are often attributed to his spirit. Zechariah 4:6 *"...not by might nor by power but by my Spirit saith the Lord"* There are at least three ideas conveyed in this phrase.

- Power or force
- Birth
- Caring

The Spirit is a person and the spirit is a force. The Hebrew word "ruwach" [roo'-akh] that is translated spirit here is also translated, "wind" and "breath". What is the wind? Is it not a force? Or the effects of a force? Sure it is. Psalms 33:6 says," *By the word of the LORD were the heavens made, and all the host of them by the breath of his mouth."* Word speaks of his orders and his commands and breath speaks of his power.

Birth and caring are implied by the word " moved" which is a prime root that means to brood- flutter Deut :32:11 – shake Jer 23:9 The mother bird imagery is brought up. The picture of a mother bird on her nest or protecting her young. The Spirit was also pictured as a dove coming down upon Jesus at his baptism. The earth was the first to," abide under his wings" Psalms 91:1 and the earth was the first to be, covered with his wings Psalms 91:4 and be *"born upon her wings*: Deut 32:11. This caring portrait of God over the planet, what is there to say about it? That God cares about this planet of all planets. This planet is not left at the whims of nature. God is watching over her. Our tiny oasis of life. so alone and unique in the universe. Although it is scared by the ravages of sin it is still the main focus of the eye of God.

It is comforting to think that we are not left to happenstance or to what the end would be according to nature. That is if the sun were to some day run out of energy or on the chance that some object from space was to collide and destroy the planet or any other way that the demise of the planet might come about. The earth is God baby and is under his care. The end, which he conceived for her, shall ultimately come to pass. Now the entrance of sin has brought and is yet to bring much destruction but we can have confidence. Revelation 21: 1" *And I saw a new heaven and a new earth: for the first heaven and the first earth were passed away; and there were no more sea."* Also Isaiah 65:17," *For behold I create new heavens and a new earth: and the former shall not be remembered nor come into mind."* What an amazing statement! What God has made he can make again and that even so much better!

First Light

The FIRST was there
At the first, at the twilight of man.
The FIRST was at the first and gave the first command.
The FIRST created the light
Light was first in his hands
The FIRST was the first to ever see.
The FIRST must be the first to have had the ability.
The FIRST was first to divide
The light from the dark.
From side to side He parted them.
The FIRST was first to name the two spheres in which we live.
The FIRST was first to say;
This one is the night and that one the day.
The FIRST was first to clock the first tick and the first tock.
First eve, first morn, first day,
The FIRST finished it all the way!

Stage 1
Day 1- Genesis 1:3-5
Creation on light
Observation and evaluation of light
Division of light from darkness
Naming of Night and Day
Noting the first day
The meaning of night and day
The basic cycle
The two modes
The 24-hour or solar day

The creation of light
"And God said, let there be light: and there was light".
This is to deal with the condition of the darkness. Just like the first thing that any of us would when we enter a dark room would be to turn the light on so God acts in a like manner. When God enters the dark universe he turns the light on. But before he could turn the light on He first has to make the light. This having to create light is a blow against those who would propose that Genesis one was a recreation of the universe as opposed to the original creation. There would be no need to create light had there been some pre Adamic world. There could not have been a world as we know it without light. Had there been such a thing God would only have needed to restart the light not create the light.

God in his mind had conceived all the roles that light would fill in the universe. It is so amazing to think of the non-existence of light and then of the existence of light. The conception of the electromagnetic spectrum and all that it entails much of which still remains a mystery to this day. The qua tom behavior, The wave theory, energy, magnetism, The dual nature of wave and particle, The information caring aspects of light. Light and optics, The heat and warmth aspects of light, The absorption and releasing of light, The constant speed of light The propagation of light through space. The connection of light to just about everything. Just trying to imagine being without it! Or to imagine the conception of light in the mind of God. What a masterstroke of God. Light is a key to how so much of the universe works and to how it is put to gather. Of all the natural forces light might be the most interesting. In a truer since this is not a natural force at all. But it is a created force. God himself is the only true natural force in the since of what has always been there and is uncreated. There is only one true natural force. All other forces are created forces.

Light is the first wonderful provision of God. God knew the need of light to our world. People are light creatures. That is we do most everything without eyes. We want to see what we are doing as opposed to feeling, hearing or smelling our way around. Sight is so essential to us. Also temperature and energy is supplied by light. The warmth to live, the energy to live, and the energy for photosynthesis.. WE have a light-based world and a light based universe. In addition, there is the metaphorical use of light. That is how we use it in our thinking . The way that light is a reflection of truth and knowledge. The contrast of light and darkness. How we take for granted and how we overlook the provision of light. May God be thanked for his conception and creation of light.

This also is the first use of the command – done formula of the statements of God..

"Let there be – and there was..." The creation of light is the first of the decrees of God. This reflections back upon the word *"created"* in verse one. The word created being a general or summary statement and "Let there be - and there was" refers to each individual act. Also it is noteworthy that in the Hebrew only the word *"said "* is used. God said light and there was light. God's saying light in this sense is also a command. The English -" let there be" expresses the command aspects of the word *"said"* as spoken by God in this context. Genesis 1:3 only has four words is Hebrew. God said light, light. The divine will is being exercised and expressed. When God said light he had conceived light and willed light into existence. Again this shows us the power and uniqueness of God. This is his mark. His M.O. What sets him apart from all others.

The observation and evaluation of light.
" And God saw the light, that it was good:"
Light exists for the first time and God saw it. God observed what he had done and evaluated what he did. This is a repeating pattern of how he did his work. This is the key of evaluations or the key of stages. Whenever God finished one aspect of his work he examined it. This examination shows that a stage had been completed. God acted in a series of stages. Every time he completes a stage he does an evaluation. There are seven examination statements verses 4,10,12, 18, 21, 25 and 31. Therefore there were seven separate stages in creation. This he begins with the creation of light and repeats to the creation of man.

Light is evaluated as *"good"*. God is creator and judge. He is also the quality control of his own work. He stands back, sees what he has done and makes an assessment.. Just as we would do. It is interesting that the word good is used. This tells us something about how God sees himself. He does not boast. He does not gloat over what he has done. Nor does he gather all the angels and say boys look at what I have done. Light is not amazing or fantastic or incredible it is just good. It was fit for what he intended. Just good seem to be an understatement but not from his perspective. Light came out to be exactly what He intended nothing more nothing less. I recall a story about Beethoven and his music; he was accused of having too many notes in his music. To this he replied that he did not put too many or too few notes. He said that he put just the amount of notes that were required. This is how God views his creation. He does just what was required. Light was just what it needed to be to fulfill the roll envisioned. Light was good.

What type of an evaluation is the word good? Is it a moral assessment? [I think it is.] A quality assessment? What standard does God go by? Whose requirement does he meet? At first this seems like a simple thing to say that something is good. We know intuitively when something is good. Some sense within us tells us. This is a function of our intuition. This is a part of our being which we have from God. When you ask the question of what does good really mean? All of a sudden I am left in a void. I do not have an adequate answer. It is beyond me to say. How do I know what good is? Who is to say? This is a deep question for the philosopher and the humanities. Here it refers to light being all that God had envisioned and all that he intended it to be. Light fit the specification that God set down for it. Light is an example of a good work of God. But this does not really tell us what good is. The answer is found in God he is the definer of what good is. It is a part of his being. God himself is the standard bearer. His intuition is the benchmark. HE holds the final say as to the quality, correctness and worth of things. God himself is the ideal of good. In the story of the rich young ruler Jesus said," there is none good but God". A practical definition then is that good means to meet God's standard. To be what God intended. Whatever meets God's standard is therefore good. God's standard is first of all himself and secondly it is what he says, His word and his commandments.

The basic cycle - the two modes.
"... and God divided the light from the darkness."
After the creation of light there is a division of light from darkness. The original light did not shine in the way God wanted it to. Light needed to be arranged into its place. The way that the light shined on the earth is what is in view here. This follows his pattern of first creating and then arraigning. Light shined upon the earth from all direction. The globe was complete lit up. This was not the final arraignment that was in store for the way light was to shine upon the earth. Therefore

the light was gathered all to one side. Presumably it was made to shine from the direction and in the intensity that the sun would on day four.

What is happening here is the creation of the basic cycle of a dark period and a light period. A light mode and a dark mode. Later he would fill the earth with creatures for each mode. The Lord gets a double usage out of the earth.. The Lord must have taken into account the length each period and the heating and cooling aspect and the angle that the light struck the earth. And the weather related aspect. The day cycle is our most common cycle. We rise, work, and then rest repeatedly. This is so sublime and so difficult to put into words. This also reminds me of how God really does control us and everything about us. The whole realm in which we live has been directly set by God.

Something should also be said about there being light without the sun. This is a common objection to Genesis. This first light and the light of the first three days were there without the Sun or any other stars. It was a supernatural light. God energized it himself until the natural source of light was created on the fourth day. The question is asked why this was done this way. For one this shows the creation of light as an entity in and of itself, the nature and behavior of light is in view. Light is separated from light sources. Now why did he wait until the fourth day to create the Sun? This does not seem logical or reasonable. Would it not seem logical to make the sun and stars along with the light? We do not know of one without the other. Once the reason for this is understood it becomes clear. This also was a part of his way of doing things. This is the key of forming and filling which was laid down in verse two. God followed a pattern of forming and then filling. First, all the things that had to do with forming were accomplished. The creation of light was a part of this. Light was formed in nature and behavior. The forming of the earth into the three different spheres of air, land and sea was part of this. Then after words he would do all the things that were a part of the filling. The

Sun and stars are a part of the filling. This is the reason for this being done on the fourth day. He formed everything and then he filled everything. He formed the light. He formed the earth. Then he filled the heavens with the light [sun and stars that is] and he filled the earth with creatures. He formed and filled, this is the logic of it. This was his plan. We simple have to catch on to what God was thinking at the time to understand it. This also keeps the earth as the focus and the earth as the central thing. God is not limited as we are. Light for three days without a sun was not a problem for him. The size issues or the effort required were not a reason to make them first. The greater mass of the Sun or the vastness of the stars did not warrant there being created first. He did not create from larger to smaller. When it came to the stars he kept to his plan. He formed and then filled! This kept the earth as the central focus. The earth was not to be overshadowed by the creation of the sun and stars. Our place is the main place that is being explained to us.

The naming of Night and Day

"And God called the light Day and the darkness He called Night."

This naming of things is another of the keys of Genesis 1. It is **the key of the name**.

The Lord specifically names five things. This is very significant that only these are named. Day, Night, Heaven, Earth and Seas. These five things again are showing us what he is doing. This is the setting up of out paradigm. These five things encompass our realm. Another way of considering this is to see it as two modes and three realms. The two modes, night and day. In addition, the three realms, land, sea and air. Looking at this mathematically you have two times three, which equals six. Therefore there are six different realms in the earth system. Then if you add in the spiritual realm there are a total of seven in all. Seven the number that God sees as the perfect number or the number of completion.

Under what conditions do we name things? When something is new. New in terms of conception or new in existence. Like a baby or the invention of some new device. This is how all five of these things were. Day and night were being invented! Along with land, sea and air. None of this had been before. Think of this also in contrast to the nothingness and the darkness that was before.

The word *"Day"* is the Hebrew – yome, and means to be hot. The energy and warmth of the Sun is in view. The word *"Night"*, layil in Hebrew holds in it a wonderful secret and a wonderful revelation. This word means to twist [away from the light]. The namer of the darkness knew the reason why the earth would go from light to dark and back to light again. That is because it was spinning in front of the light. Now how is it that such word is used to name the night? The answer is that this word if from God. How very well this shows that this is divine revelation. How scientifically accurate is the word night! Also how befitting it is that such a secret is held in the word night. Because the study of the night sky has been the revealer of so many truths to humankind.

What a great disservice the church did by adopting the Ptolemy model as doctrine without examining their own writings in more detail. The great rift that developed between science and the scriptures. The rejection of the bible as inaccurate. The writing off of the scriptures as a guide. The struggle of the church to accept scientific fact. The development of scientific theory regarding earth history and life independent of the bible. The mindset that was present throughout the intellectual world when Darwin proposed his ideas would not have been there.

All of this may have been avoided. The evolutionary revolution would not have taken place. The great falling away and turning away from the bible would not have happened.

That the earth was spinning which of course we cannot tell by standing on it show that the bible stands apart from all other writings. It is not like the myths of other ancient religions nor does it have the errors of early scientific investigation. What is says holds true through all ages. See also Job 26:7, which add also to the scientific accuracy of the bible which says. *"He stretched out the north over the empty place, and hangeth the earth upon nothing"*. The spin of the earth and the floating of the earth in space are both revealed in the bible. Science and the church may have walked hand in hand in unity and harmony down through the ages one backing up the other. The scriptures would not have lost their prominent place. The question has been asked of what is in a name. Well in this case a great deal was in the naming of "Night"

The 24-hour day
"And the evening and the morning were the first day"
The passing of the first day is noted. This goes along with the idea of the basic cycle and the division of light from darkness. The passing of every day of the first week is noted in Genesis chapter one. The word "Day" is linked to a single morning and evening cycle. This is repeated each and every time with each and every day. Verses, 5, 8,13,19,23 and 31. The meaning of day is clear and specific. The meaning of the word day is limited to a single solar cycle of one evening and one morning in Genesis chapter one. Nothing else is accurate. Nothing else is said here. In chapter two verse four the word day is used to speak of the entire creation week. The whole creation event is in view. This different usage is clear from the context of the verse. Also there is no mentioning of the evening and morning phrase in 2:4, this allows 2:4 to speak of a longer period of time but not in chapter one. Science may not agree. Okay so let science be in disagreement but do not twist this word into meaning some other period. God has said that he did his work is six days. Seven when you throw in the Sabbath. This show that science just misses the mark in there evaluation of the

history of the cosmos and the earth. The believer in Genesis must simply say they are wrong. There is no other choice to make. Accept the revealed word or accept scientific interpretation. Let the battle line be clear. The believer to be true to the position of the scripture must hold all scientific evidence contrary to this as suspect. A long time is only a necessity for the naturalist who must have so much of it for the processes which he proposes work. Or to those who would like to reconcile science and Genesis while accepting the science as accurate. God has no time limitations. He worked in the chosen increments of days as a pattern for us to follow. He followed the same cycle that he was creating for man. It all could have been done a moment had he so wished.

The noting of the first day is a moment beautiful beyond words. God witnessed it. God marked it. A thousand verses of poetry could not express it! Day is a creation of God. All the aspects of our earth day have been specifically defined by God. Also remember that there were no other planets around that could have had a day cycle of their own at that time. This is the setting up of the most basis aspect of our paradigm, of our way of living. A time of work and a time of rest. Now we have something to count And something to follow. Our life is lived one day at a time. It was the first of a cycle that will continue forever.

There are two or three other side issues that are related to the twenty-four hour day cycle that are worth of mentioning. First the meaning of the words of Genesis 1 is at stake. These words have real and specific meaning. To make them cover some long duration of time is an abuse of the language. It would make these words mean nothing. Moreover, the writer of Genesis would not know what he was saying. Second the extent of the glory and power of God is lessened. If God did all this in literal days then this shows the extent of his power as opposed to some much longer period. Just as the lifting of a load faster shows a greater amount of force was available than lifting the same load over a longer time. Finally evolution cannot be true if they were literal twenty-four hour days. Time and chance are the basic ingredients of evolution and this is not enough of it for evolution to take place.

Backward Rain

Upon the shinning, tranquil and endless sea
The pressure of the first air came to be.
Where once only water touched space
The atmosphere has suddenly taken its place.
Oxygen and nitrogen in just the right amount
To empower the breath of ever creature had been
taken into account.
The proper mixture of mass, gas, and space
To provide the need of the human race.
Then a pair of cupped omnipotent hands reached
down and scooped
Out a measure.
Upward outward heavenward they rise
To take their place above the shimmering skies.
No time before or since has been
That the rain fell backward!
What a mystery it is to us is made
That water one existed above the shade.
The write of Genesis had a story to tell
Of how the waters were there before they fell.
Sea and sky are made
This is the second day of God at his trade.

2 - Genesis 1:6-8
Introduction
Study of the word "firmament"
The water canopy
The absence of the waters
The naming of the realm
The order of events

Introduction
"And God said," Let there be a firmament in the midst of the waters"
The next task, which God had, was to create what we now call the atmosphere. First came the basic cycle, which is Night and Day and secondly came the realm of Heaven, or the atmosphere. At the end of day one space and sea touched. There was nothing in between. It is noteworthy that an air space is being created in between two bodies of water. *"In the midst"* that is in the middle of the waters. This situation does not exist today. We today understand the importance of the air. What it means for the respiration, pollination, flight, temperature, wind, protection from harmful rays and weather. So also did the writer of Genesis recognize the distinction between outer space and the atmosphere and the importance of it to the earth and even that there was once a great body of water above the air.

The study of the word "firmament"
The word "firmament in the King James version follows the tradition of the Latin Vulgate. The root idea there is of something that is fixed, secure, solid or sure. This does not remind us of what we know of our gaseous atmosphere. I believe that the Latin choice of this word refers to the Ptolemy's model of the universe, which the church held, from the third century and through the Middle Ages. That is that the sphere of heaven were solid and fixed not the empty or gaseous void, which we know them to be today. It seems that

the translators of the Authorized Version in the sixteenth century also may have leaned toward this position. But this is found to be both scientifically and textually inaccurate. The original Hebrew is a much better source of reference and gives us a much better understand of what is going on here. The Hebrew word is
"raqiya" strong s # 7549 is used in Genesis 1:6 and means properly," an expanse " that is - Open area. Take into consideration Genesis 1: 20 which says the fowl would fly in the " open firmament of heaven". Also consider that " raqiya" come from the root " raqua" Strong's # 7554. This word is a prime root which means," to pound the earth " and by implication to overlay. [with sheets of metal] This word comes from the goldsmith's trade. When the gold was beaten into thin sheet and used to overlay an object such as the ark of the covenant or an Idol this is the word that was used. The earth was therefore overlade with the atmosphere. This is the correct implication of the Hebrew The location of the atmosphere is more in view that the substance of it was. The creation of this space does note that there was a difference between it and what was there before and between it and what lay on the other side of the waters above. The expanse is noted as being separate and open. This special space between the waters was being set up. This speaks of our atmosphere with all its layers and different gases in their various amounts.

The water canopy
"And let it divide the waters from the waters"
This expanse was situated in between two great bodies of water. Portions of the original waters were lifted up and above the atmosphere. Again the Lords pattern of first creating and them arranging is seen. In addition, this is the second part of how each day has two major events. He created the atmosphere and divided the waters these are the two great events of day two. Which result in the creation of the realm of Heaven

First just the fact that the scripture notes the existence of this is important. That something was there that is not now. Neither Moses, the author of Genesis, nor modern man could have known of it.. Not only does the Bible have knowledge of the future that no other book has but also it has knowledge of the past that no other book has. The scripture is true forwards and backwards. This adds to the point that Genesis is literal and actual history of true things and events. Why would any ancient account of creation include such an idea? Why would they have any clue as to the advantages of such a layer of water above the earth? There is no example of this anywhere in the known universe to draw this idea from. There is no other creation account that mentions any such idea. This is a powerful argument in favor of Genesis. There is no logic or reason for the inclusion of such a thing except be a revelation from God. Modern scientific investigation has no clue about. This water canopy is never a part of their models of the condition of the pre mortal earth. Neither is the original water covered globe for that matter. Naturalism would love to propose such life favorable conditions! In addition, there is no real evidence or legitimate theory of how the earth got its oceans or its atmosphere anyway only a list of possible guesses. I wonder how they would have tried to explain it if the canopy had remained to this day? God tells us of what does not exist anymore. Why? Because that is the way it actually was. The earth was so designed so that our type of life could thrive under the protection of this water canopy. How very different the earth was [and is] from all the other planets. Moreover, how much more different it was in view of this water canopy.

Also today we know that such a situation is plausible because the waters could have been placed out there at such a distance and in such a form that they would stay there because of the weak gravity at that distance. Ancient people did not have such knowledge. The volume of water must have been very great. First because this moving of the waters is one-half of the events of the day. The atmosphere created and the waters moved being the two events of the day. Secondly as witnessed by the flood of Noah when these waters rained down for forty days and nights and helped completely flood the earth. Creation scientists believe that this water canopy was very beneficial to life on earth. That it protected life far better from the harmful aspect of the Suns radiation.. That the climate would be uniform around the globe and that the seasons would have less temperature variations. In addition, the atmosphere pressure would be greater because of the weight of the water. The longevity of the antediluvian people is accounted for because of these early conditions along with the giant ism of creatures found in the fossil record. It was truly a different world. A great amount of power was necessary to raise those waters. We see today how powerful a rocket must be to get a few tons of materials into space. Can you imagine the energy to raise an ocean? God show his omnipotence repeatedly in this chapter. He separates the waters, raises the continent, creates and places the sun and moon and stars in their places! The displays of power are on a scale that only his divine hand could perform. Many times in scripture these works are alluded to as the standard and example of what only he can do.

The absence of the waters
First, the now missing waters are testimony of the judgment of God. Someone might ask where are they now? The waters, the one great continent along with the four great rivers, Eden all gone. That they are not there is strong witness to the fact that God has judged the world. What the Lord can put up he can bring down. The waterless sky is a witness to us. This speaks of moral responsible and privileges. The condition of creation is connected to man, his behavior and his relationship with God.. The waters were there for our blessing and brought down for our judgment. God knows how to bless us and how

to curse us. There is the anthropic principle that shows how the universe favors life for man. In addition, there is a moral principle in which the universe is affected because of man's sin. God is there in the background causing these things because of his moral nature.

There is a practical side to this, which indicates that I affect my universe through my moral choices. We might call it the principle of moral effect. We affect families, our society, even nature and ourselves by our morals!. God has woven the world together in this manner. Yet man is often the last to come to recognize this and accept his moral responsible and to see it for how extensive it really is. This load of moral responsible is heavy and assume but it is also something that we can have working to our advantage when we accept it and live under its terms.

This is principle, which the naturalist fails to find in nature. His paradigm does not allow for such a thing. Nature is amoral to them. Only a personal being can have morality, which is to make moral judgments.. In many ways it is true of nature [that it does not make moral choices] but not in all ways. For example the absence of the water canopy and the curse upon the productivity of the ground in Genesis 3 are examples of this moral principle at work. There are others such as the giving and withholding of rain etc.

The Naming of the realm
"And God called the firmament Heaven."
This phrase brings us back to the key of the name. And to the purpose of the events of verses six through eight. That is the creation of the realm of the sky, Heaven as God calls it. The second of the three day of forming. This is the Hebrew, "Shamayim", dual of an unused singular "shameh". meaning to be lofty. First this word refers to the area immediately around and above us, the space in between the seas and the water canopy. Later the same word is used to speak of the space above the water canopy, what we call outer space. It is

interesting that this is a "dual" word. This word names two spaces so I suppose this is why it is a dual. The Biblical view of the physical heavens then has two divisions. Earth's atmosphere being one and then what lies beyond. The moon planets stars and galaxies are all grouped into one. The Hebrew word for water "mayim" is also a dual. It speaks of the two vast bodies of water that existed then. One below and one above the atmosphere.

God named the realm. This is very significant. This is what he was about doing. This was His purpose. Bring this into being. Creating and naming this realm. This was a new thing!. The first of the three realms of Heaven, Earth and Seas. We today are so inundated with the evolutionary processes that it is difficult for us to conceptualize this as an idea of God. God thought it up. God produced it and God named it. God hold the patent on it A planet does not have to have an atmosphere. God added this to the earth.

The order of events.
Why was the atmosphere next on the list? Is there anything to learn because of this order?. Was Heaven made next because it was more important? Was the Lord working from outside to inside or top to bottom? Is there any reason whatever for the approach he used ? I believe that there is. God again can be seen to be acting with reason, logic and efficient. Even thou he has all power he still worked according to a well thought out plan.

First the atmosphere needed to be in place before the events of day three. The plant life of day three would need both the protection of the water canopy and the presence of the air to survive and to function. Secondly a portion of the original waters would have to be moved in order for the dry land to appear. The sea level would be lowered and the continent would not have to be raised so high to reach the location the God had envisioned for them. Had he moved the waters later he would have to re due the land and oceans. God created the earth with the total volume of water for the waters on both sides of the air in verse two. More water was not needed. Part of it just needed to be moved. This goes along with how he first created the raw materials and then went about arranging them. The Heaven was made first because it was necessary and efficient. The work was easer that way. When we catch the grand plan then we are more able to understand why each step took place as it did.

Half a day's work

It was half a day's work, that's all it took,
The water flowed greatly and the land shook.
So great was the roar and the volume of the sound,
That even the MAKER took note and had to tone it down.
A greater movement of mass in such a short time has never been,
Yet he was not weakened by it at all He could do it again and again.
IT was in the dark that the last drop slipped by
Valley, plain and hill are exposed, even the mountains high.
The rising of the continents the sinking of the seas,
Were all a part of the grand plan God had for you and me.
Sky and sea were not meant alone to be our earthly home,
So up arose the dust and dirt on which we now presently roam.

Stages 2 completed&3
Day 3 Genesis 1:9-13
Formation of Land and Sea
Naming Earth and Seas
Evaluation of Earth and Sea
Creation of Vegetation
Replication and multiplication of vegetation
Evaluation of vegetation & summery of first three days

The formation of land and sea
"And God said, Let the waters under the Heaven be gathered together into one place, and let the dry land appear: and it was so."
1: 9

THE two events of this day are, one, the revealing the dry land and the covering of the land with vegetation.
The Lord works on a gargantuan scale. By one command the final two realms are molded into shape. The sea floor drops away, the waters follow, the dry land is uncovered. Earth and atmosphere touch for the first time. This takes half of day three. Then the land is covered again this time with the plant kingdom. Note that there is a distinction made between the two great bodies of water. Only the waters under the Heaven are addressed. This is keeping in mind the previous work of separating the waters. It seems that the sea floor sank more than the continent arose. This is suggested by the wording, "*let the waters be gathered*". Either way the results would have been the same.
There was one ocean and one continent in the beginning. The waters are gathered into "one place". Modern geology confirms that the continents as we know them now were once all together. At least in this one aspect there could be some agreement. The bible records the continent being one and its being broken up. [see Gen 10:25]

The crust of the earth now goes through a major change. Whatever the original shape of the sea floor was in verse two is now changed. I can only speculate that the original floor was uniform or close to uniform in depth. There may have been some change in the sea floor due to the weight of the waters that were removed on day two. The reduction in pressure may have caused a rise in the sea floor? . We can only speculate in what the verse two sea floor conditions was like. One thing we do know from this is that this movement of day three would remain evident in the record of the rocks layers. This support the cataclysmic view of the geology and stands against the uniformitarian model of earth history. The sea floor dropped hundreds or even thousands of feet and that over and around the entire globe. And all of this in about twelve hours' time.

What an amazing event it would have been to witness!. To have been suspended above in a balloon or some type of aircraft. To see the great rush of falling water, the hear sound, to feel the rumble of the descending crust. To watch the rising of the continent. What cracking, braking, folding and moving of the crust there must have been. The incalculable energy that it took. The planet being molded in the hands of God. Nothing like this has ever been witnessed by man. How in awe we are of a volcanic eruption. When a comparatively small amount of ground moving. Much less trying to comprehend a whole continent and the whole ocean on the move all at once. Every act of God during creation week is beyond any scale that the earth experienced since that time. Our imagination is strained to comprehend these things.

Again the principle of God's methodology is seen. That of first making the raw material and then placing them into their proper arrangement.

The naming of the final two realms
"And God called the dry land Earth: and the gathering to gather of the waters called He Seas:"

Earth, Hebrew, Erets, Strong's # 776 means "to be firm." This name is given in contrast to the liquid state of the water. The uncovering of it was a new thing. Again the key of the name in invoked and the intention of God is seen. From this word we get the name of our entire planet. This is probably because man is a land creature. The planet is named in connection to man to his planet. But the point being a whole new realm is formed. A surface area of dry land. The basis for the habitat of air breathing creatures. The air and land now meet. The great continent, the original land mass.

This dry area is a conception of God. Again it is difficult for us to recognize that this is so. That dry land is a purposeful thing of the mind of God. He has is mind a whole store of creatures who will live upon, within and off the land. An entire ecosystem will be based upon this .He might easily have left it as a water only planet or a water and air planet. All creatures could have fit into just those two but he did not. Man is especially connected to the land. This is the final step in the setting up of our paradigm.

The waters also are now named. Seas he calls them. This is the Hebrew, Yam, strong"s # 3220 from a unused root meaning to roar. The sound of the water is where this name comes from. First I think of the noise that there must have been of the waters rushing off the continent.

What a grant sound it must have made! Think of Niagara falls and then multiply that by a million. Secondly the seashore and the sound of the breakers come to mind. This is one of the things we notice when we visit the ocean. The Heaven is named in relation to its location, the Earth from its solidness and the Seas from the noise it makes.

The evaluation of Earth and Seas
" and God saw that it was good"
This marks the completion of the three realms. Light is evaluated in verse four then there is not another evaluation until verse ten.[Day and Night were not a part of the first evaluation.] In addition, the Atmosphere has not been assessed after its creation.. All these things are taken together at one time and in one group. Remember these are the named things. Day, Night, Heaven, Earth and Seas. The two modes and the three realms. All the environs of our planet. The areas of operation of all creatures. These things are seen as separate and complete from what was left to come. The first stage is the stage of light. The second is the stage of the formation of sea, land and air, our geosphere.. These are the two grand movements thus far.

The day cycle does not all ways co inside with the stages. Day three and six each have two stages completed upon them. Day one, four and five complete one stage each and day two covers only a partial stage. What is there to learn from this? God kept several things in mind as he worked. He knew how long he was going to work. One week that is. He knew what he wanted to accomplish. He knew the cycle he wanted to follow. He went about it in the most necessary, efficient and logical way possible. Things were kept in groups according to an overall plan. God continues to follow the day cycle. Even when no stage was completed as in day two or even when a day completed two stages as in day three and six. The reasons for the cycles falling as they did were not the same as the reasons for the stages being completed as they were.

The creation of vegetation
"And God said, let the earth bring forth grass, and herb yielding seed, and the fruit tree yielding fruit"

The vegetation is the final step of the three days of formation and the final step in the preparatory phase of the earth. The earth was not completely formed until it was covered with vegetation. Plants therefore are seen in there connection to the earth, as a part of it, more than there connection to the animals , as a part of living things. They are not seen as living forms of life as the animals are. The word life not used until verse twenty where the waters are filled with" the moving creature that hath life" That plants are nonliving things from God's perspective is important for us to grasp both because of when they were created and what service they perform in the overall scheme of things. They were created and intended as provisions and substance. Even though they have many chemical and biological functions, complex cell machinery, and even reproduction yet all this is made for the living creatures that were to come not as part of them. The ability to move, to have self-locomotion, seems to be what sets living things apart from nonliving things. The definition of what is alive, what is not has been a difficult line for science to draw, and maybe things would become clearer if the biblical standard of self-locomotion were used instead of trying to define it on the chemical level.

Plants were created at this particular point because First - the protective atmosphere was in place. Second - the earth was now exposed and therefore ready to be covered. Third - because they needed to be in place before the animals were created. Many make the argument that this is an illogical order of things because the sun did not exist yet. There are at least three reason why this in not so. The most important reason it that God sees the plants as a part of the formation process and not a part of the filling process. He sees the heavenly bodies as part of the filling process and not of the forming. Which relates back to the key of methodology found in verse two but besides this there are other good reasons. First there was light present from its original creation on day one. Therefore, there was light present for the plant to function. Secondly could not the plants last a short while without the sun? Sure they could. This adds to the argument for literal twenty-four hour day periods during creation week. Next it is important that there had to be a command for the plants to come forth. They could not come forth on their own accord. The soil could not produce them through any attributes of its own. Only plants can produce plants and God made the first plants. The answer to age-old question of which came first the chicken or the egg is solved here. God, who is eternal, made the first chicken, which then made the egg. Evolution fails at this point. Plants from inert soil cannot be demonstrated. Seeds grow is soil but soil cannot produce seeds only plants can do that. The gap between soil and seed cannot be bridged. Just as the gap between the nothing of Gen 1:1 and something of Gen 1:2 It is God who fills in the gaps.

The vegetable kingdom in Genesis is seen to be separate from the animal kingdom. It is both different from the earth, or soil, and from the animals. The plants were meant to be food for all creatures. Plants do not die for the same reasons that people do. There is no moral or judgmental connection to their death. They were and are meant to be consumed. This is their purpose. Even though they do have many life like functions as we have yet God has categorized them differently from us. They were made for us and we were not made from them as the Darwinian view would propose things.

Bible believers need to have a separate classification system that evolutionists do. Plants are divided biblically into three broad groups. Grasses, herbs and trees. All plants should be fit into one of these. Also there is a division between seed, herb and fruit. There are three basic types of plants, which produce three types of food. There are several groups of three's in Genesis. For example we have seen the creation of the three realms, land, sea and air and now we see three groups of plants. O whole study could be done on the number three and groups of three in Genesis.

Replication and multiplication of vegetation
" the herb yielding seed after his kind; and the tree yielding fruit, whose seed was in itself, upon the earth"

This is the first mention of reproduction and replication. The importance, significance and the marvel of the seed. The great storehouse of information, which they contain. The ability to germinate ,replicate and multiply. God made the plants, which produce the seed, which reproduces the plants. Again God has filled in the gap between the inert soil and the complex seed. Anything else is just a fairy tale. This is how it is this is how it works. There is no natural mechanism to explain the seed. To the naturalist a great puzzle and mystery and a strain on the brain. However, for the believer a marvel of the ingenuity of our magnificent God. We have seen the strength of God in separating the waters and in lowering the sea floor but now we see God in his engineering and intellectuals skills. To have conceived of the niche which every plant would fill. To write the code. To lie in down in the genetic language. TO make the entire cell machinery of the first plants. Just one is more complicated that can figure out. Yet he did it over and over countless times and all in a few hours! The many things, which he must have kept in mind as he did it. How each plant would respond to light, temperature ,soil, water conditions, and other plants and to animals. Which would dominate what environment. Which would grow fast, slow. Which would be long or short lived. A thousand million different aspects all taken into account and woven together into one ecosystem.

For many reasons this should be the end of evolution. The central tenant of evolution is that all life forms are related through the inheritance and change of traits down through the ages. The seed and how it works shows that this is not so. Each plant type or, kind is unique unto itself. There are variations of traits within each kind and these variations show up according to the rules that Gregory Mendel first discovered. Traits are not acquired and then passed along as

Darwin proposed. The information and the design of each aspect of every species are already there in the code. The only option that evolution had to take after the confirmation of genetics is that through chance mutation of the DNA coding [of which they cannot explain its origination anyway] all the various life forms have come about. The precision and delicacy of the code makes this very unlikely at best and in reality impossible. It would be like a person was copying a book, say a dictionary, but when he was done because of his many mistakes it was now a science fiction novel. In reality this in just not going to happen. The only relation all things have unto each other is that all have the same creator who used similar engineering to produce each individual. The tree of life as evolution proposed it should to be changed. There is no flow from one ancient ancestor down to man. Each species is its own tree which originated from God and remains within a set of limited already present variations.

Plants are God's wonderful system of provisions. Just as the preparations of the geosphere with its anthropic form for the benefit of man so is the plant kingdom. One seed can produce one plant, which will produce many hundreds of seeds, which can produce many hundreds of plants, which will produce many thousands of seed. In this multiplication system there is provisions for all. The survival of the plant and the supply of those creatures that would live among and off of them.

That the scripture mentions this fact of their creation is noteworthy. Plants are different from the soil that they grow in, vastly different. They compose a major stage of creation by themselves. There is a great gap between them and the inert soil that cannot be explained by any natural process. There engineering, chemistry, metabolism, genetics and reproduction all set them apart from the soil. The soil is the habitat and substances for the plant. In addition, there is a system for the soil to be renewed because when the plant dies it returns to the soil. We see in Genesis a building up process. First of the planet itself. The air, water and land are laid down then the soil for the plant and the plant for the animal and the animal for the man and the man for God.

The earth and heavenly bodies do not regenerate they were created from God's hand and remain much the same except for the effect of their running down. Plants do regenerate and that according to their kind, after what sort they are. The seed was in itself. The plant makes its own seed, which in turn makes another plant. This system originated with God and is recognized in scripture. In addition, I might mention that DNA is the servant of the plant and not the other way around. It is the macro system of the plant in relation to earth's ecosystem that is to be kept is view and not the micro system of DNA and the mechanisms that make the plant function. Some evolutionists like to look at it as DNA is somehow trying to insure its own survival by all that take place in reproduction but this is not the case. DNA has no purpose in and of itself. DNA is not trying to do anything. It does not think or have knowledge or will. This is an error of reductionism. That is how science goes about to reduce everything unto it is most basic parts and then reason its purpose from there. Without faith in God many wrong conclusions are drawn about the meaning and purpose of things in nature. Science is good at answering the question of "what" but is bad at answering the question of "why". Plants and animals are more than cell machinery. Even though they are made up of and due function because of this machinery.

Evaluation of vegetation and summery of days one through three
" and God saw that it was good"
As before the evaluation marks the end of a stage. The plants were good. They were what he intended. The earth is now covered in greenery. What a difference from the uncovered and barren landscape of a few hours ago or the submersed continent of the day before. Leaves move in the breeze. Trees stand tall. Shrubs and grass cover the open area. Flowers

bloom. This is the third great movement of creation week. This far we have the stages of: light, the stage of the geosphere, [the five named things] and the plant stage. These three bring to an end the first half of the week. In three grand movements we have come from nothing to a planet that is ready for inhabitants. The habitat is completed. There were three thing pointed out in verse two that God would go about to rectify. He went about to light, form and fill the earth. The first two are now completed. There was "darkness "and God made light and the earth was "without form" and God has now brought it into form. These two can be seen as preparatory movements. Plants are seen as part of this preparatory movement also. They are part of the forming of the earth. All that is left is to fill the earth with its intended creatures.

For a moment the earth existed alone, complete and empty. Nothing else in the universe, by itself. No sun, moon or stars. No other planets and no creatures, Nothing. Only the light of day one accompanied it.. The night sky was completely black, not the tiniest ray of light shown from anywhere. How dark the night must have been. How would it have been to look up into a completely black night sky and see nothing at all or to try and get around in that darkness. In addition, what different perspective we would have of ourselves had he left it that way. Many things that we know now would be unknowable.

Our tiny sphere, which may not be at the center of the universe but is central to the universe, sat alone, FIRST and them the rest of the universe was added around it. God built the earth and then he worked outward. This order is according to his purpose and his intention. This is according to what was most important to him and to what he was doing. The earth where the grand drama of man and redemption would be acted out. The earth where God will eventually make his home. This order is connected to our sense of meaning. The way we view ourselves is linked to the way we view our planet. What meaning is added to us! Our place was made first. It is first in importance in the entire universe. How the naturalistic view has robbed the earth and man of their meaning. From the principle thing to nothing and meaninglessness. Again without God there is nothing but meaningless.

How about to ponder for a moment that earth and plants are older that the heavenly bodies?

To our evolutionized minds a strange thing but not to God. To him it was logical and meaningful to complete the acts of formation first. The time and timing aspects of things are much different with him than they are to naturalistic science. Time is not important to God but timing is. We see time as some sort of creator itself. We have been sold the line that if you only have enough of it that anything is possible. This is wrong thinking. An eternity passed before Genesis 1:1 and there was still nothing. Time in and of itself can do nothing. It is God that has acted and not time that has acted in bringing the universe into being and making the earth.

Lights

Odd it seems God and his plans,
For they could not have been carried out by human hands.
That sun moon and stars did appear
Four days after the first of the spheres.
For HIM this was a logical choice,
Because to do anything HE needs only HIS voice.
For three days the heavens stood empty,
Then came the fourth and there were stars aplenty.
Of particular interest were the two great lights,
one would rule the day and the other the nights.
The four seasons also would be caused by his mighty,
Winter, spring, summer and fall all for our delight.
Also a part of what HE had in mind
is how it is that we would keep the time.
To the day is added the month and the year,
Their purpose told in the scripture it's all right here.
Thus were the heavens filled as incredible as it seems,
the earth being the central focus in the universes scheme.

Day 4 Genesis 1:14-19
A change of perspective
Why the sun, moon and stars were made the fourth day?
Why light the first three days?
The cyclic aspect of heavenly bodies
The light source aspects of the heavenly bodies

"And God said, let there be lights in the firmament of the heaven"
A change of perspective
The earth is created before the stars and the stars are not created until the fourth day. All this calls for a change in our perspective of things. This is the key of the earth. The centrality of the earth is again emphasized. The earth itself is to be seen as one of the keys to understanding Genesis one. The sun moon and stars are seen in there relation and purpose toward the earth and not the other way around. They are made and placed to serve the earth. We tend to look as the solar system as the sun and its planet's as companions. But this is inaccurate from the biblical view. Even the term solar system is misleading. We do not live in a sun system. Biblically we live in an earth based system. The size, placement and gravitational effect of the sun are not to be confused with its place in the scheme of things. The earth is still to be viewed as the body of primary importance. This also applies to the entire universe.

When we found out the comparative size and placement of things in the universe such as the sun's mass and location and then the size of the galaxy and the finally that there are many millions of galaxies then we also changed our view of planet and ourselves in response to these discoveries. But this should not have been done. Everything out there has been placed there with the earth and us in mind. Mass does not equal importance. Nor does the number of stars make them more significant that the earth. Mankind has drifted far from the world view of Genesis. The fall of man, the separation, the withdrawal of God, the loss of Eden, man going at it on his own since then and evolutionary theory has all clouded our perspective. Along with the loss of our soul has been the loss of our identity. We no longer know where our planet or we fit into the grand scheme of things. The meaning of ourselves is linked to how we view our planet and its place in the universe. The centrality of the earth needs to be regained along with the view of ourselves as being made in the image of God. This place and all the surroundings are here for our benefit. The sun is a light provision.

Why the sun, moon and stars were made the fourth day?

This is an important question to ponder. Why did God wait until the fourth day to make them? Why not go ahead and make them on the first day along with the creation of light? Then at lest you would not have to explain how there could be three days of light without any light sources. Many are stumped and bewildered at this order of things. This seems an impossible logical hurdle to jump through. Or is it? If we believe that God is logical and reasonable them there must be some reason or reasons that he acted in this order.

First why he waited the three days before he made the heavenly bodies. Chiefly it is because he followed a twofold process in creation. It is the process of forming and then filling. This is in keeping with the issues he noted in verse two, The key of methodology. God went about to light, form and then fill in response to the things that he noted of the original conditions of the earth being without form, void and dark. The celestial bodies are a part of the filling. So there creation is delayed until after the works of forming were complete. The creation of light on day one is part of the forming but the creation of light sources is a part of the filling. Creation week can be divided into these two great movements. The first three are days of forming and the second three are days of filling. He formed light and the earth and then he filled the heavens with lights and the earth with creatures. God followed a repeating pattern according to his overall scheme. That is after the days of forming God went back to the first thing that he formed and then filled that thing and worked his way back down the line. This would them be a repeated order of; [forming] light, atmosphere, seas and land. And then [filling] heavenly bodies, fowl, fish, animals. [light, air, seas, land] So this actual does fit what he did! The heavenly bodies being created on the fourth day is done in accordance with this grand two fold plan. This is the logic of it. This really is a key to understanding the chapter. We allow ourselves to be stumped by the size differences or by the occurrence of a miracle of supernatural light. But for God acting supernaturally is "natural" and this order was reasonable. This is in perfect keeping with the context of the chapter. In the study of Genesis one the church has failed to catch on to what God was up to. God followed certain ground rules and a plan. The meaning of verse two as a guiding factor has not been recognized and without it much has remained a mystery. The conditions that were called out in that verse were not happenstance. They were full of purpose concerning what God was about to do. We do not have enough faith in our own doctrine of the divine message that the bible is. Also we miss these things because of what we have been taught. Our teachers spent their time telling us the gap theory or the day age theory or trying to make some excuse for why the scripture does not line up with modern geology and we spent our time doing the same. These ideas do not really satisfy our minds. Nor do they do justice to the grandeur and glory of God. They are weak and miss leading. Really they are compromises offered as allowance for evolution and for the appeasement of modern geology. The great mind of God is missed. His grand twofold plan; His orderly arrangement of events. His timing of things. The heavenly bodies being created on the fourth day are not odd or unexplainable but they are a master stroke of God. His timing is perfect and his logic for doing so is impeccable.

Why light the first three days?

Another good question is why have any light for the first three days at all if it was not a part of the plan for the light sources until the fourth? God could have created light as a thing without making it shine for three days... Why three days of light without light sources? The answer is in keeping with the key of cycles. Both following the day cycle and setting up the week cycle. The reason lies in God's overall intention and scheme. God could not have followed the day cycle or the week cycle without the light of the first three days. God intended to work for one week. This week cycle was and is very important to him. This is a chosen increment of time. He could have created it all in one day or one moment had he chosen to. Had he not lit the first three days then there would not have been seven days of creation and the number and order that he envision for things would have been thrown off . The week would have been only four days long. And this is not what he intended. The Seven day week was what he was after. Creating and following these cycles was one of the paramount aspects of creation. Important enough to have three days lit by super natural light so that he could have literally the first three days. The day cycle and the week cycle are prime chosen increments which God set up and worked according to. This lends support to the literal view of actual twenty four hour days and one seven day week of creation just as we know a day and a week to be. God followed both the forming and filling process and the chosen cycles of the day and the week. These are two of his overall and overriding parameters of creation.

The cyclic aspect of heavenly bodies.

"... to divide the day from the night; and let them be for signs and for seasons and for days and for years."

The writer of Genesis understands the roles that these things play in their relationship to the earth. The text is very accurate, day and night, because the earth is spinning in front of the sun, the seasons, because of the angle of the earth to the sun, the year, because of the rotation of the earth around the sun. Then there is the moon that governs the monthly cycle and causes the tides. There purposes are earthward. It is what they do for us on the earth that has placed them where they are and as they are. This invokes the key of the earth and the anthropic principle.

The scripture speaks of this in terms of the purpose that these things play as regarding the earth. This also means that the writer of Genesis knew of these things in terms of what caused what. He knew the cause of day and night, the seasons and the year. We have been told both purpose and the cause in Genesis.

One of the flaws of evolutionary philosophy is that the idea of purpose is missing because purpose suggests a creator. You may discover the grandest design but then according to evolution you must explain it without suggesting that came about on purpose. This is a very tall order in the face of all the wonders of nature. Volumes are written trying to explain why a design was not designed. Design and purpose are philosophically taboo if you are an atheist. Reductionism and meaninglessness are the only options you have. You can break everything down into its parts but can never say that it was put to gather that way on purpose.

The divine inspiration of Genesis can be seen in what is says about the purpose and cause of the celestial bodies. Thousands of years passed before man discovered this scientifically. But the writer of Genesis had already said it long ago. No other ancient writing can compare with Genesis. No other writing in the world can compare to the bible. We should give credit where credit is due. The sun and moon are not gods and receive no worship in Genesis. They are just as we know them to be today. The heavenly bodies were never mythologized in the bible. While the Egyptians worshiped the sun as a God Moses sat down in the desert and pinned it down that the sun was a creation of God and that God was beyond the universe. How could Moses who was raised and taught as an Egyptian have made such leaps in his thinking? Was Moses the first modern scientist? Did he study astronomy in the desert? No. God revealed himself to him and then told him about it. What Moses wrote was a complete shift of paradigms. A totally different way of viewing the world than what he had been taught.

The word "sign" has many uses is scripture, here it means to mark. This has to do with time keeping. We mark the time by observing the heavenly bodies. We would have no means of keeping it without them. The universe is like a gigantic clock. There is the day night hand then the moon hand then the planet hand the star hand and finally the galaxy hand. This is how God has provided for us to keep time. Eternity was timeless but the universe is full of time events and markers. Again the idea of purpose is invoked. It's not just that we can use them to keep time because of happenstance or coincidence. But because this purpose was divinely ordered that we can do so. The clocklike aspects are on purpose.

      The key of cycles can also be seen here. God set up the basic cycle of day and night in 1:4. Now to it is added the month, seasons and year.

In 1:4 *God divided the light from the darkness.* What he did there through supernatural means is now take over by the heavenly bodies. Night and Day are divided, or caused, by the arrangement between the earth and the sun.

There are seven different naturally ordained cycles. The Day / Night cycle, the month, winter, spring, summer, fall and the year. Then they start over. This corresponds with the week which has seven days and then starts over. By the way the week is the only cycle that is not connected to any naturally accruing event. IN this it is unique being established solely on the word of God. Seven is the number of completion in scripture and eight is the number if new beginnings. Maybe we should rename the days of the week with meanings that correspond to these seven cycles instead of how they are now named after pagan gods? But we would have to keep Saturday as the Sabbath. Since the week was ordained in honor of creation it would be reasonable for the meaning of the names of the days to be connected to creation also.

The light source aspects of the heavenly bodies
"And let them be for lights in the firmament of the heaven to give light upon the earth: and it was so."
As with other things in nature God gets more than just one usage out of the celestial bodies. We have already seen that they function as a clock and cause the cycles. Now they are spoken of as light sources. This purpose also is stated. "to give light upon the earth. Light provides for us energy and information.
" These lights that he speaks of in general are then more specifically defined in verse sixteen as" two great lights; the greater light to rule the day and the lesser light to rule the night;" They are defined according to their brightness and in relation to their effect upon the earth. The sun provides the day light and the moon is the main source of what light there is at night. The stars also are mentioned. "he made the stars also" Again we might invoke the key of the earth for it is there role in relation the earth that is the cause of the sun and moon being as they are.

Three categories of light in all, sun light, moon light and, star light. Just as there are three categories of plants; grasses, herbs and trees. What wonderful variation of light God has created. First there is the blinding sun that blocks out all others during its rule of the day. Then the much softer moon light with its varying phases and monthly cycle and finally the much fainter stars with all the mysteries they hold.

As with the other days of creation there are two grand movements or events on day four. The first is the creation of the light source verse 16. The second is the placement of the light sources in their locations verse 17.

"And God set them in the firmament of the heaven"

We are not given any detail as to what set where in relation to each other. We are told that they are in the firmament of heaven or outer space as we understand it to be today. No orbital information is given. Nothing is said about what was going about what. We were told earlier that the earth was rotating, this was found in the study of the word Night but this is the only information about any of the motions of the heavenly bodies that we received. Neither are we given any size, motion, and speed or distance information.

Why not? The only textual reason that I can think of has to do with the overall context of chapter one. God has described creation to us with a broad brush. Many details have been left out. He used broad and large categories which are inclusive of multitudes of details. Had he given us all the detail Genesis would have been a book of thousands and thousands of volumes. Just as he did not give detail of how genetics works in relation to the plants so he has not given the details of size, speed, and locations of the heavenly bodies. God has not told us everything but what has told us is true accurate and fitting. Non information is much different than wrong information. There are other passages of scripture that give further information of the topic found here. For example Job 26:7 and Isaiah 40:22

"he made the stars also"

Again this is a short phrase to convey such a mass and a multitude of things. Here also there are other passages that give more detail such as Isaiah 40:26 and Job 22:12 It is to be understood that the rest of the stars were made at the same time as the sun. First because they are categorically the same as the sun. Secondly because it is impossible that they could have been made any time earlier. Light itself was not made until day one and could not exist without light. Also the universe was in darkness in 1:2 so they could not have been in existence then likewise.

Many take issue with the scripture over the issue of the stars the creation account and the distances that they are from the earth. I have heard many possible explanations to this issue but none that I could place any confidence in. To these I would like to offer one of my own for us all to consider.

One thing that I have seen is studying Genesis one in that God followed a pattern of first making and them arranging thing. For example he made the raw material of the earth in verse two and then he went about arranging the earth. Then with the sun and moon in verse sixteen the same pattern is followed. He made the two great light and then set them in the heavens. I propose that he did the same with the stars. In Isaiah 42:5 it says *Thus saith God the LORD, he that created the heavens, and stretched them out..."* What if he created them some place astronomically close the earth and then he moved them from there out to their present locations. Which is in accordance to his pattern of working. This then could be a reason that we can receive their light even though they are now at such great distances. In the stretching them out they left a trail of light from where they were to where they are now. Also since the stars have also been made with the earth in mind it only makes sense that God would make there light to already be shining on the earth. From the intent stand point in would make no sense to make stars that it would take millions of years to pass before they could be seen from the earth. Also from the intent stand point this would show that God has not tried to deceive people by placing some scientific stumbling block in front of modern man. It is man that has failed to understand God's plans that has caused there to be an issue.

Ocean Life

How pure, how clean how clear, how vast
Were the waters that had yet to see the mast
And much more amazing to grasp
Is that through them not a single fin had past
Not a flipper or a fin
Not a shark with its evil grin
Not the ray nor the ell
Not a boat or a keel
And above the oceans blue
Not a single fowl ever flew
The waters were empty
But in a few short hours they would be life aplenty
Not that everything in them had died
Only it was that they had not yet arrived
Then again the voice was heard, echoing
Across the waters and falling upon the
Waves and crashing upon the shore
Like the great breakers
Be filled, be filled, be filled!
The waters stirred!
Life broke upon them like a storm
Life fell like the rain
Life arose like the flood
Life streamed like a river
He fills the liquid realm just as he will the space
A thousand million wonders each with glory on their face
Fish large and fish small
Fish long and fish tall
Fish short and fish fat
Fish round and fish flat

Types to numerable to mention
Created by him in a moment's decision
And how could I even begin to account
All the floating, wiggling, or crawling things
That are in a fount
The microscopic life by the billion
That have the water as their pavilion
And last but not least of all the
Great whales that roam the watery halls
The water itself which is so unique to
The earth were filled with life five days
After their birth

Wings

Of the many abilities that God has designed,
None of them seems to be so sublime.
As the wing.
The bird had been given the right,
To pass through the air by day or by night.

What a freedom to come and to go as they please,
And all of this with the greatest of ease.
The idea of flight is envied by man,
But for him it's not yet a part of God's plan.
This for man awaits him for a future day,
In a kingdom that's not that far away.
How is it that man could possible imagine,
That the marvelous design just happened to happen?
No not even in a million billion years,
[the truth of this even Darwin fears}
Could it possibly have come about,
That on some crawling, running or jumping creature wings could suddenly begin to sprout.
What a multitude of coincidental changes it would take,
For the first airfoil to have taken its shape.
How could anything develop the feather?
It's light weight and strength they are so clever.
I suppose that it must have been due to the weather!
But to those of us with faith and common sense,
This is all a bunch if non sense.
For God has given the eagle the ability to soar,
Just as he made the sun to shine and the lion to roar.
The wing with all the knowledge it suggests:

*The knowledge of lift
And the knowledge of drag.
I believe that evolution has just been had.
The knowledge of atmospheric pressure,
Should prick the heart of any professor.
Take a moment sit back let it all come into focus,
That I'm not speaking of hocus pocus.
But of a God who is so great and so vast,
That at the sound of his voice it came to pass.*

Stage 5
Day 5 Genesis 1:20-23
Water creatures
Life? What is it
Air creatures – Birds
The divisions of sea and air creatures
The word "kind"

Gen 1:20
"And God said, Let the waters bring forth abundantly the moving creature that hath life, and fowl that may fly above the earth in the open firmament of heaven"

Having filled the heavens God now returns his attention back toward what is left to do upon the earth. As with the other days there are two grand movements on this day also. They are the creation of the sea creatures and the air creatures. Again the brush is very broad. The creatures are created according to the realm that they would inhabit. This reflects back upon the key of names. The three realms that God created and named are now going to be filled. The water and air are probably filled before the land because man is a land creature and God was saving him for last. It is not that water and air creatures as any less advanced than land animals.

The division of creatures according to their level of proposed advancement or complexity is a false evolutionary concept. God gifted every creature with the specific attributes that they would need to live in there realm and fulfill their nitch in nature. How can you compare for example the sonar of the dolphin with the echolocation of the bat? Both use the same means but one is within the water and the other the air. Is one more or less advanced that the other? How can you say which should have come before the other? Or the eye of a crab with the eye of an eagle? The crab has no need to see as far as the eagle so its eye is made differently. Complexity should not be an issue. It is fitness that is the issue. Each is perfectly fit for its environment. To say that one existed before the other because of complexity is first a philosophic error based upon the assumption of evolution and secondly impossible to determine. Is complexity just the number of pieces that a mechanism has? No it is not. This is a wrong basis of evaluation.

God in creation has explored every possible avenue. This is why so much has been said in so few words. He has made creatures in a multitude of forms with countless varying attributes and abilities. It is his endless creative mind and ability, His vast engineering skills, His artistry, His music and his poetry. Nature is the display of his hand. Evolution is a total denial of it. A poison that deceive the soul.

We are not given any subdivision for these creatures. They are all lumped together as simple sea creatures or winged fowl. To explain them all here would be too lengthily. God saw fit for it to be enough to just say that they also are of his creation.

This is also very different from the evolutionary order of life. Creatures are created biblically according to their realm of sea, air; or land and below man. Not according to simplicity, complexity or some imagined evolutionary ancestry. The categories are fixed and unchanging. Fish started as fish, birds as birds, animals as animals, man as man each according to its " kind ". There is none of the changing that evolution proposes. Of course this means that much is wrong with the science of paleontology. The bones do not tell us the relationships of one creature to another as they say they due. The whole evolutionary scheme should be thrown out and a new one started that reflect the created order. Creatures should be categorized first according to their realm. That is are they land, sea, or air creatures. Next according to their mode. Mode means day or night creatures. Then according to their means of locomotion. Other divisions, classification or groupings could then be made after these.

Life? What is it?
" the moving creature that hath life"
What is life? How do we define it? This has been a difficult thing to define. Does metabolism or chemistry represent life? Does reproduction represent life? In the scripture the life is in the blood. See Lev 17:11 But what is the most basic defining aspect of life? From this phrase above we might get the idea of self-locomotion to define what has life and what does not. This ability to move being the most basic aspect of what separates life from non-life. Three things define or relate to life. They are breath, blood and motion. Of these the most basic is that of motion. If something can move on its own then it sure to be alive. Is not this the way we often check a creature to see if it is alive or not? Sure it is. We touch it or poke at it to see if it will move. If it does not then we are sure it is dead.

Plants are not said to have life nor do they have blood. In addition, plants do not have the ability to move on their own. The wind may blow them and other creatures may carry them around but they cannot move on their own. Nor do they take in oxygen. Plants are a category of their own. Separate from soil but also separate from living creatures. Plants receive their nourishment differently. Plants breathe differently. This is again a place where evolution blurs the lines of the uniqueness of the categories of the created order of things.

Locomotion may also reflect self-will is some way and therefore be representative of life. The ability to move co insides with the will to move.

It is no coincidence that these two words are used together.

In speaking about what life is I must also comment on how life arose. Scientists know that there still is no adequate explanation for how life arose. How can they hold to evolution as being true without it being able to answer this most basic question of how life arose in the first place? You cannot call upon survival of the fittest when there were no creatures to compete with one another. You cannot call on gene mutation when there were no genes to mutate. There is no principle in nature, which can change the inorganic into the organic. Alternatively, which can take non-life and turn it into life. When it comes to the origin of the first life naturalism draws a total blank. The situation is just as Louis Pasteur said, "life only come from life." The scriptures declare that all life on this planet arose from God. The God who has eternal life has passed life on to us and all other creatures. This would be in line with what we know from nature.

The complete evolutionary concept of life is wrong. This idea that life somehow through happenstance came into being and then proceeded incrementally to develop and spread and to improve itself. The logical trail that this proposes it wrong. Nature speaks against it.

Air creatures - Birds

", and fowl that may fly above the earth in the open firmament of heaven"

After the waters are filled God turns his attention toward the sky. The creatures of each realm are all so unique to their environment. Birds are so different from sea creatures. The requirements of air living and water living being so much different. The mode of movement being the principle difference. God does not mention the difference of being able to breathe under water as opposed to in the air, which is what we, might think of as the principle difference. Indeed there are sea creatures that get along just fine breathing air the whale, dolphin and walruses for example. But the difference in the mode of movement is mentioned. The ability to move is mentioned of sea creatures but it is undefined. Therefore, this can include swimming, crawling and wiggling about. The propulsion systems of sea creatures are many and varied. However, when it comes to the air all fly. God knows what the most basic differences are. This is also a part of his wonderful thinking. We may even call this a key. That is the key of movement. This is the principle that God used is creating the creatures. It is the filling of the realms that he has in mind. All the unique aspects of the birds are made with the intention of them filling their intended realm. The size, weight, strength, and design requirements are all in accordance to the intention and purpose of making creatures fit for life in the sky. How much more sense this make that to try to explain how this happened by evolutionary processes. Are we really to believe that flight could come about by accidental mutation with all the requirements that there are for it? Really the same complication exist for all modes of movements it is just that the contrast is more obvious when it comes to flying. Thinking humanity should be embarrassed by the idea. Such is the delusion of evolution.

The divisions of sea and air creatures
"And God created great whales, and every living creature that moveth, which the waters brought forth abundantly after there kind, and every winged fowl after his kind"
The water creatures are divided into two categories. The whales and then everything else. The birds are all in just one group. A couple of other distinguishing factors are noted. Of the whales there size is referred to as great. The word whale is a general term referring to any creature of monstrous size. Of the birds there wings are noted. It is the secret design of the wing that enables them to fly. The writer of Genesis knows that there is something special about the wing. How do you say enough about the wing or the feather? Or the understanding of lift, drag and thrust? What a contrast also there is between the whale and the bird. What different creatures they are. The massive lumbering whale as opposed to the light and agile bird. Yet both a conception of God perfectly fit for their realm. One roams the deep and the other the air. And we marvel as them both. There is perfection and completeness is both.

One of the problems with trying to explain the world though naturalist means is how to you explain all the perfection. In evolution perfection is denied as only a mental perception
Another problem is that there is no synthesis that should be found in nature if evolution were true. If evolution were true we would see creatures as a blended and blurred together without distinctive categories. However, what you find in nature are creatures perfectly fit for their realms and there nitch in the world in fixed categories. This brings up the problems of the lack of transitional forms in the fossil record and the lack of living creatures that are in transition. They just do not exist now in the present or in anything that we know of in the past. It is not just that there are a few little missing parts in the proposed chain of evolution. There are giant gaping holes and vast chasms missing between things. For example the gap between non-life and life. Or the gap between plants and animals. Or between the two basic kinds of cells.

The word "kind"
"after there kind and after his kind"
God tells us that each creature reproduced a like creature or a like species. This is important and accurate. Kind produces kind and nothing more. Kind does not produce some other kind. This is right in line with what we know of genetics. Nothing can be produced that was not there in the first place. As with the plants this takes us back to which came first the chicken or the egg. God created the originals and they reproduce after their kind. A believer does not have the problem trying to explain how creatures arose by happenstance or how they change as the evolutionists does. We are told how they remain the same and how they originated in the first place. The picture is clear, complete and reasonable.

Evolution is always going about trying to explain how creatures can be so fit for their environment by naturalistic means, chiefly by natural selection. The formula is:

genetic mutation X reproduction = all creatures

This formula is totally inadequate to explain what we see in nature. The genes do not function as the formula proposes. This idea of what genes do is totally opposite of how they function in nature. The genes function to reproduce the " kind" not to change the kind. The mutation of the genetic code brings harm and dysfunction. Harm and dysfunction would reduce a species chance of reproduction not increase it. The chance of a chance mutation being beneficial is so remote as to be for all practical purposes impossible. The very word mutation shows the weakness of this process. When something is a mutation it means that it is no longer what it was meant to be. What a great long shot it is that any gene mutation could bring about a favorable condition. Much less the millions upon millions of positive changes that it would have taken for creatures to evolve into what we see today. Not to mention the fact that the creatures would also have to remain viable and dominate while they are going through the process of changing. In addition, if this were true then how is it that creatures can remain in their form? If genes are such a powerful force for change then what is it that keeps a species together? Evolution seems to want to have its cake and eat it too. Genes cannot be the means of both. Let the matter be clear genes function to reproduce creatures as they are not to change them.

Enigma

Last but not the least of the days
Is the one in which the man was made.
Man the creature separate from all others,
Man who is so cruel to his brothers.
Man the creature, who failed to do his best,
Man the creature God longs to bless.
Man the creature that lost his first estate,
Man the creature whose rib became his mate.
Man the creature that would name the rest,
Man the creature that failed the test.
Man the creature that lived a dream,
Man the creature gave into Satan's scheme.
Man the creature that once had it all,
Man the creature so ruined by the fall.
Man the creature, who knows not from whence he came,
Man the creature now full of sin and shame.
Man the creature which still bears the mark,
Man the creature brought alive by a divine spark.
Man the creature that shares the image of the Lord,
But hardly can you tell it because it's been so scored.
Man the creature the chief of the earth,
Man the creature that God offers a new birth.
Man the creature the riddle of it all,
Man the creature will walk through heavens halls.
Man the creature the maker of this rhyme,
Man the creature that lives for all time.

Stages 6&7
Day 6 Genesis 1: 24-31
The creation on land creatures
The creation of man
Man – a creature patterned after God
The name of Man
The dominion of Man
Man and his gender.
Blessing and instructions given to man
The final evaluation

The sixth day follows the same now familiar pattern as the other days. There are two major events on each of the six days of creation. The two major events are different from the stages. The stages do not always co inside with the days. For example, day six has two stages within it whereas day two is only part of a stage. Each day has two major events but each day does not have two stages. In addition, each completed stage has three broad categories.

Now in the case of day six the two major events are each a stage of creation in and of themselves. These events and stages in this case, are the creation of land animals and the creation of man. The three division of the land animals are; cattle, creeping things and beasts of the earth. The stage of man also has three divisions because man is a three-fold creature made of body soul and spirit.

In addition, it is to be noted that this is the filling of the third and final realm of the earth. This reflects back upon the key of names. God created and named five things. Day, Night, [the basic cycle] Heaven, Seas and Earth, [the three realms] this day brings us to the filling of the final realm and the completion of the earth.

The creation on land creatures

"And God said let the earth bring forth the living creature after his kind, cattle, and creeping thing, and beast of the earth after his kind: and it was so."

The terms used are again broad and all-inclusive yet there are also some distinctions to be made. How do we define what is biblically in the group called cattle? Of course, the common cow would come to mind. Any earth creature of great size may be the general idea here. Most of the time this Hebrew word is translated as cattle but it is also used in Job 40:15-19 of a much greater creature:

"Behold now behemoth, which I made with thee; he eateth grass as an ox .Lo now, his strength is in his loins, and his force is in the navel of his belly. He moveth his tail like a cedar: the sinews of his stones are wrapped together. His bones are as strong pieces of brass; his bones are like bars of iron. He is the chief of the ways of God: he that made him can make his sword to approach unto him."

May be if it eats grass as a cow does then it would be included in the category of cattle? [Everything ate plants at that time, see verse 30] In addition, we can see that this word is also broad enough to include the dinosaurs. Behemoth certainly describes something to us that is far greater than a cow. Behemoth may have been a better translation that the word cattle because the author is pointing out to us the extremes and the range of things. As when concerning the water creatures, the word whale is used. The idea is that the greatest creature is pointed out then the rest are referred to in general terms. Just as when one person would describe the airplane to another, he might speak of a 747 as opposed to a Cessna. The AV translators probably used the word cattle because this is the most often used meaning of the word and because of their lack of awareness of other creatures such as the dinosaurs. Just as the Hebrew word for whale in verse 21 is a broad, enough term to include the leviathan of Job 41 God made great whale *and everything in-between,* might be the idea.

Creeping thing is a little easier to understand. This would refer to everything in the insect world. These different levels and these separations of things are very important to our understanding. The insect world stands apart from the cattle world. Each is a separate order of creation. Yet each shares the same realm, the same environment. The elephant and the ant each has their place and their role to play.

Not only do we need to recognize the separation of the levels of creation but also we need to acknowledge the interrelations between the levels of creation. For example how the insects pollinate the plants. Or how some insect live off other creatures. These symbiotic relations are also the creations of God. It is not reasonable to think that these relations evolved together. That the bee just happened to mutate in such a way as to make use of the nectar of the plants that just happened to mutate in a way that would be beneficial to the bee is really far beyond anything that you could reasonably expect from such a principle of happenstance that evolution proposes itself to be. Moreover, for this to have happened repeatedly in the countless way that we find in nature. What an assault on sound reason the theory of evolution makes! Logic, reason and probability all say that this is impossible to have happened of its own accord. The strict and well defined division and separation of the levels of creatures and yet the interdependence of the same is a stroke of the divine hand. The different level as distinctive as they are, are yet interwoven and interdependent. When we can admit that that there was a plan in all of this then we can understand the logic of it. God had planned for the harmony that we see between the flower and the bee. The one was in fact created with the other in mind. No other explanation will do. The scientific community often accuses the believer of a lack of realism in what they preach but I believe that the same charge might be made against them. They, the scientific community should accept the fact that these things could not have happened without there being a plan in mind. This of course would cause evolution to crumble, the lack of any such plan being one of its chief corner stones.

Finally the phrase *"beast of the earth,"* would include everything else. The word beast is a very general word that means any land animal. Again as with the water, creatures there are far too many of them with far too much variation to be more specific and to so would have made Genesis to long a book. God is a God of variety. There is more than one-way to skin a cat and there is more than one way to make a land animal. He made one creature this way and then another that way. Yea He went on as it were a rampage of making creatures! With His vast and endless mind, with his ceaseless power and energy he has spewed out the various different forms. God is a great creating, manufacturing and constructing machine! Each with its unique behavior, each with its niche to fill. For every nook and cranny, for every hole, every space, every level God has planned devised, conceived and created. Until he purpose to fill the earth was complete. What we see in creation is that God has acted as God. This is the explanation for the wonders that are all around us.

"and God saw that it was good." As noted before this phrase marks the completion of the stage. When God finished a stage he evaluated it. He checked his work as he went along. This is the first half of the sixth day and the completion of the sixth stage. Each stage has received the same evaluation. Each is equal in quality and measure. The repetition of this phrase is not just being redundant but a true examination and assessment of its condition. The work of the Lord is always perfect. Every stage received his best attention. One not any better or worse that the other.

The creation of man

"And God said, Let us make man is our own image, after our likeness: and let them have dominion over the fish of the sea, and over the fowl of the air, and over the cattle, and over all the earth, and over every creeping thing that creepeth upon the earth."

First, I would like to note the distinctiveness of man. The fact of man as a stage of creation in and of himself. Each stage of creation is separate and unique from the other stages. Naturalism tends to focus on the likenesses instead of the differences. However when you compare the stages to one another the stark contrasts and differences stand out. It is no accident that the other land animals are evaluated independently of man. God evaluated the stages because they were individual, separate, unique and complete. Each stage stands alone. Each stage is a category of its own. Just as separate as light is from the earth and the earth is from plants and plants are from stars and stars are from fish and birds and fish and birds are from land creatures so is man from the rest. Man as a stage of creation in and of himself is not to be understated. Man is separate from the rest. Man is not just one land creature among a multitude of others. This is seen first by the fact that he is a stage unto himself.

Man – a creature patterned after God

Secondly man can be seen to be different by comparing the phrases: " *after his kind or after there kind*" as spoken of the creation of the plants and the animals with the phrase "*in our image, after our likeness*" as spoken of man. The plants and the animals are each new conception of God. They follow no previous pattern. They are never before existed creations. They reproduce according to the new kind that they are. That is they follow their own new pattern. Nothing existed that God could compare them to. Their kind had just been created and were made according to their kind. But when it came to man he is patterned "*after our likeness.*" In Man's case, he does follow an already existent form. Man has been patterned after God. The comparison and contrast of these phrases shows this difference. As you study through the chapter the repeating phrase, after their kind stands out and you even come to expect it to be said again whenever a new kind is formed however, when you come to man this phrase is replaced by the phrase, *after our likeness*.

As earth, creatures we share many things in common with other earth creatures yet we also bear something that none of the rest have. It is this image of God. If you ask the question, who or what are we most like. Or what creature are we most closely related too? The amazing answer is that we are most like God himself! You see the monkey is not our closest relative. The monkey is not even in the same stage of creation that we are. We are not to look unto the monkey to discover who we are but unto God. The more we find out about him the more we find out about who we are supposed to be. The more we come to conform ourselves into his image the more we rediscover what we were intended to be. There is a line of reasoning that says the way for us to understand ourselves better is to understand where we have come from. If we have come from the monkey then one set of behavior might be expected but if have been created in Gods image then a completely different set of behavior and expectation. This uniqueness and difference of man created in the image of God cannot be overstated. Many volumes can be written and have been written expounding upon this idea. Another thing about this is that is if man is made in the image of God then the opposite is also true. God is like we are. God is a person as we are. With a like nature with like behavior, a being like unto ourselves. When we discover our true selves then we discover what God is like. Now in extent he is far greater yet in nature he is the same. Our nature is a copy of his nature, our form a copy of his form. We are a miniature of God. In the same way that we might produce a scale model of a car or some building project so are we toward God. The fall has marred his image in us, history has obscured it and we have gone about to redefine ourselves yet this is the grand truth that is revealed in the creation account. The great difference between man and the rest of creation then is our similitude to God. We are a reflection of him. We resemble him. We have been patterned after him. We bear his own personal image.

The name of Man
The term "Man" is the generic term for what we are. Man is named in connection to his place, to his world. The Hebrew word is " aphar" Strong's number 6083. Which comes from the prime root of 6080, which is translated as the word *"dust"* in Genesis 2:7 Man is made of the dust of the ground and is also named after it. This is in accordance with man's place. Psa 115:*16 The heaven, even the heavens, are the LORD's: but the earth hath he given to the children of men."* The earth is our domain. The earth is the place where men rule. One might think that man would be named after his being in the image of God because of how significant that is but the Lord has named us more in connection with our place. Our earthliness and our being the prince of this place are connected to who we are. Man is the chief creature of the earth. We are named in accordance with our domain. Every creature has been given its notch, its area, its habituate, but man has been given the whole of the place. The world truly is our playground. We inhabit, explore, investigate, manage, enjoy and affect every area. Even men who do not know God still enjoy the world. We love this place. WE soar in its air, surf upon its waves, run across its land, climb its mountains, swim its seas, explore its forests and dig in its soil. Yea far too often people have much more appreciation for the world that they do for its creator.

Also, it is befitting that man is named after his world because the world was made with man in mind. This reminds me of the anthropic principle, how that everything seems to have been made to benefit man. This is because it was.

How then are we to think of ourselves? How should we think about selves in light of these three facts? That we are made in the image of God, and that this place was made for us and that we are the rulers here. Special aren't we! What a creature of worth and importance we are. We are at the pinnacle of the physical creation. How wonderful and good God as treated us. How privileged we are. Also how responsible we are. As goes the man so goes his world all is affected by what we do.

The dominion of Man
"and let them have dominion over the fish of the sea, and over the fowl of the air, and over the cattle, and over all the earth, and over every creeping thing that creepeth upon the earth." Verse 26

The universe along with man's place and position in it are all a direct result of this image and likeness of God that we bear. As God has a domain, so man has a domain. This is right and reasonable, the one follows the other. For man to not have had some area of dominion would be to have made him less that the image of God. The earth is the domain of man. This also may include the entire material universe. Man is the ruler of the earth. Set so by God in accordance with his likeness to God. Adam and Eve were created King and Queen, the head of this whole creation. Man is made of the highest possible order, After the Chief himself. God could not do anything greater or higher that to make us in his own image. Man is a creature of the highest possible order, worth and dignity.

Man did not struggle to get to where he is as evolution proposes. He was made to fit, made to rule. Man has not fought his way to the top. He was made at the top and has fallen down. What a very different conception of whom we are and how we got be where we are. Man has not evolved but devolved.

The extent of man's dominion is over all the earth. The three created realms that were made and named are each mentioned. The sea, air and land are the extent of his realm.

Man and his gender.

Gen 1:27 *So God created man in his own image, in the image of God created he him; male and female created he them.*

This is the first mention of gender in the chapter. This was not brought up concerning other creatures even though they must have been created male and female so that they might reproduce after their kind yet here the issue is raised of man. Why? The reason may have to do with the image of God. The implication here is that the Godhead consists of more than one gender. The male without the female in an incomplete picture of God. Likewise, the female without the male is an incomplete picture of God. The image of God in man is not found only in what a male is by himself but also the female. Not only this but also the family as a unit is a picture of and a copy of the image of God. Man not only as an individual but also man as a family unit is the image of God. The family is a picture of God. God is masculine, feminine and God is a family. In verse 26 God said let, us make man in our image after our likeness. To whom is he speaking? He is speaking to himself. That is to the rest of the Godhead. The Father, Spirit and the Son. This is the Godhead family. Much can be learned of ourselves and of God from this idea. The family is the creation and image of God. The family is a sacred picture of God. It is to be held as an ideal. When we break apart the family, we are diminishing the image of God. When the family lives together in happiness and harmony we are portraying God is his fullest extent. This is why the things that mean the most to us have to do with our families. God exists eternally as a triune harmonious family unit with male, female and childlike aspects. We think of male and female being primarily for the purpose of reproduction but there is a part of it that goes beyond that. A part that is deeper and more profound. His image is incomplete as masculine alone.

The arising of the genders according to evolution is very difficult to explain. When did it begin? How did it develop? How could such a complicated thing evolve? The different and yet complementary organs and behaviors of sexuality all somehow arising through chance. As far as evolution goes why are there only these two, male and female that is? Why did nature not explore many other varieties of sexual or reproductive forms? Why not some creature that requires three or even more sexes? Why did only two evolve? What was to limit it? If there was some advantage of there being two sexes, them why not even more advantage by three? The answer again lies in the fact that there was a plan behind it all. The ideal of gender and sexuality comes from God and is of God.

Blessing and instructions given to man

Gen 1:28 *"And God blessed them, and God said unto them, Be fruitful, and multiply, and replenish the earth, and subdue it: and have dominion over the fish of the sea, and over the fowl of the air, and over every living thing that moveth upon the earth. Gen 1:29 And God said, Behold, I have given you every herb bearing seed, which is upon the face of all the earth, and every tree, in the which is the fruit of a tree yielding seed; to you it shall be for meat. Gen 1:30 And to every beast of the earth, and to every fowl of the air, and to everything that creepeth upon the earth, wherein there is life, I have given every green herb for meat: and it was so."*

As in verse 22 where the water and air creatures are blessed by God so, he does the same with man. Man is blessed to reproduce, to make more men. The word replenish is the same as the word fill in verse 22. This goes all the way back to God's intentions toward the earth as first noted in verse two. This is not to be regarded as the refilling of some place that was once full and then became empty. God desired for there to be many men. The being a multitude of us is a sign of his blessing. Secondly, man is blessed as the ruler of the earth. These are the two primary blessing. Man is to fill and to rule the earth.

These are the relationships of man toward the earth. Now rarely do we ever think of ourselves in terms of these things but even without acknowledging them, we certainly have and do carry them out in a way that is quite natural to us. This world is ours to fill and ours to rule. How wonderfully God has blessed us!

Then God gave us instruction concerning what to eat in verse 29&30. The plants were made to provide the food for all earth's creatures. As noted before plants are not said to have life as animals do. They have no feelings or thoughts. They cannot know pain or death. Fruit, seeds and herbs were meant to be our food. Consuming them is the original created order of things. The earth was created a peaceful place. Creature did not prey upon creature. Hunting and fishing were not practiced. Man was a planter and a farmer. This was a great guiding principle for man and animals to follow at the time. This was the original economy of the earth.

The ocean creatures are left out of this instruction concerning eating. This is because there realm, the water realm was different that the land. There is a silence concerning how they went about eating in the oceans. But we may infer from the instruction given to the others that they also had a way of feeding that did not involve hunting and killing each other.

The final evaluation

Gen 1:31 *"And God saw everything that he had made, and, behold, it was very good. And the evening and the morning were the sixth day."*

This evaluation differs from the rest in that here the superlative is used. *".. And behold it was very good"*. The others were good but this one very good. The man and the woman were now there and the rest complete. The whole was evaluated as greater that the parts. The completeness brought it up to a higher level than was conveyed in the individual stages. The grand stage for the drama of the ages was now set. This was certainly a poignant moment. As we might do when that first new car rolls off the assembly line, or when some grant building project has come to completion so the Lord does here. The other stages were evaluated separately and individually but here all is examined together, as a whole and complete assembly. God stood before his perfect creation. Everything was in order; everything was complete, all functioning as on grand harmonious whole. A brand spanking new planet and universe. Very different from six days before! All this without any error or sin.

God was pleased. This is the first hint of emotion and excitement that we see in God. A hint, a glimpse of his personality peeks through. God experiences pleasure and satisfaction with his work as we do. IT was not just a technical evaluation but it had personal meaning to God. Maybe we could even see his love for the place here. IT was very good he was pleased with how it turned out and loved it. Our tiny speck of the universe draws the attention of the creator. The affairs here are of utmost concern to him.

The other planets are never brought to completion as the earth was. We do not know if at some point God intends to make them inhabitable or if they are there just for how they serve the earth. The earth stands in stark contrast to the rest of the planets and the rest of the universe as far as we know. A tiny oasis of life. A place like no other. This is a great riddle for the naturalist to contemplate. The unlikeness of the earth to the rest and the causes for this being the chief question of science. The believers answer to this is that this is of the ordination of God; God has planned the universe so. Philosophically if life arose here of its own accord then it is reasonable that it could have also done so elsewhere. But search as we may we have not found this to be the case anywhere else. The rest has been found to be very inhospitable toward life. Philosophically we really should expect to find the little green men out there somewhere, or at least little green something or another's. Not finding them is not because the universe is such a big place but because they are not there to be found. If the world has life because of the plan of God then it is reasonable that this could be the only place in the material universe so endowed by its creator. There is no necessity that it should be found elsewhere. The overall philosophy if scripture speaks against this idea of life out there somewhere else. For example in Romans, chapter eight twenty two the whole universe is affected by the action of man's sin. This moral principle affects the whole universe. It does not seem reasonable for our sins to effect creatures or people out there on some other planet so the implication is that they are not there. It may seem strange to the modern mindset that we are here alone but certainly, it is true to what we know.

For only a short time would the perfect original conditions remain. The Harmony would soon be broken. IT did not take long at all for creation to get its first scratch and its first stain. This make those first few days or years that much more special.

There is somewhat of mystery concerning where the devil was at this time and when he fell. We know that he had fallen by Genesis 3. It seems from what God has said here that at least at this moment he is not present. Had he fallen yet? We really are not told. The subject matter of Genesis 1 is man and the material universe not the spiritual world. My suspicion is that he had already fallen sometime before the material creation. It seems too much of a coincidence that he would have fallen at or around the time of the material creation. The story of Satan is not a part of the creation story. There is nothing inherent in Genesis that demands that the fall of Satan was somehow a part of what had taken place here. No, God is telling us our story, the story of the original one and thus far only material creation. The spiritual world is not being addressed in the evaluation. Therefore it may be that that Satan had already fallen or maybe not. The angels are part of a separate spiritual realm and a separate creation of God, which took place before the material creation.

Day 7
Summery and Sabbath day
A completed place
The seventh day

Gen 2:1-3 Thus the heavens and the earth were finished, and all the host of them. And on the seventh day God ended his work which he had made; and he rested on the seventh day from all his work which he had made. And God blessed the seventh day, and sanctified it: because that in it he had rested from all his work which God created and made.

A completed place
The universe is not an evolving place but a finished place. God ended his work. He ended it because it was complete, finished. The claims of science regarding the ongoing evolution of things should be brought into question. The universe in not in an ongoing process of creating or recreating itself. No knew life forms are arising. The same goes for the celestial arena, no new stars are forming as science proposes. There is a process of running down that is taking place according to the laws of entropy and thermodynamics but nothing new is being created.
The concept of creation is so different of that of evolution. Evolution does not know how it began, how it sustains itself or where it is going. It has no conception of completion or perfection. The universe is an ever-changing thing with no rhyme or reason. Chaos and order are of equal value, importance and significance. Really, chaos is what you would expect to find at the end of the day. But a universe that has been created by an omnipotent hand suggests a different circumstance. A place of completion, order, design and perfection. A place that has been arranged with a purpose in mind. Which concept fits what we see?

The seventh day
Gen 2:1-3" Thus the heavens and the earth were finished, and all the host of them. And on the seventh day God ended his work which he had made; and he rested on the seventh day from all his work which he had made. And God blessed the seventh day, and sanctified it: because that in it he had rested from all his work which God created and made."

God finished on the seventh day, rested on the seventh day and blessed the seventh day. We might expect a blessing upon the animals and upon man but why upon a day itself? God has made the seventh day special. He sanctified it. That is it has been set apart from the others. The reason given is," because in it he had rested from all his work which he created and made" God created and practiced a work rest cycle. Six days of work one day of rest just as he would later incorporate into the Jewish nation. The seventh day commemorates his resting. Before it was a part of the law, it was a part of the created order. Adam and Eve no doubt followed this same weekly pattern. The seventh day was set aside to them. The ideas of rejuvenation and enjoyment may lie at the core of the purposes for the seventh day. Now God was not tired, he did not rest for himself but for us. That is to set the pattern to create the weekly cycle for man to follow.

The seventh day was set aside to honor God. It is a direct witness to the creation event. It is a memorial to the creation. From nothing, from emptiness to a complete grand vast, complex, precise edifice which we behold from day to day and night to night From no creatures to countless multitudes of creatures. From no stars to innumerable stars. From no earth to a pristine perfect earth. From no man to the man created in his image, to Adam and his wife Eve. None ever worked harder, none ever worked smarter, none ever worked faster, none ever worked with more power, none ever put more thought into it, none have designed better, none executed a plan like The Lord our creator has and did. None brought forth life as he has. None filled the sky and the sea and the land as he has. None, none, none.

The fact that he made this a weekly memorial should be paid special attention to. We remember great human events by creating special days such as the fourth of July where we in America celebrate our gaining political independence. Or days such as Memorial Day where we remember those who have died for our country. These are yearly remembrances. But the seventh say, the Sabbath is a weekly remembrance. This sets it apart from all others. It's a greater more significant event that any that any human has had any part in. It seems that it would be too much and too often is we celebrated our independence once a week. I am sure that we would grow tired of that, and rightly so. But with creation this is not so. Also with Honoring God this is not so. It is ever bit worth of a weekly remembrance.

Another question that can be asked is this. Why did God create in seven days as opposed to some other increment of time? As we study the number seven throughout the rest of the scriptures we see that God often uses this increment of time in his dealing with man. For example there were seven days before the rain came. Seven days the people of Israel marched around Jericho, seven good years and seven bad years in Josephs dream, seven years tribulation and many others. There may be a relation of this to the nature of God. In rev 4:5 *"there were seven lamps burning before the throne which are the seven Spirits of God"* Somehow God is complete in these seven spirits. Because of this sevenfold aspect of God he may have created after this same pattern which is found in himself. the seven day creation and the seven day week may be a reflection of himself. This first period of seven days is foundational to many other groups and usages of the increment of sevens which are found throughout the rest of scripture.

How to Build a Universe Extras
Keys of Genesis

One other way of looking at Genesis chapter one is to consider the chapter though a series of key ideas or guiding principles. Even though these have been mentioned throughout the text, still it is helpful to see them all together. This will assist us in seeing the grand way God went about his work following these guiding ideas. Also this can help us wrap our minds around the whole chapter.

The two guiding keys of methodology
Forming and Filling

This key was brought out in verse two during the study of the words "without form and void". Forming and filling is the antithesis of without form and void. We can see in this that God thought about how he was going to go about his work. The original unfinished state or raw material state of the earth set it up to be formed and filled. This forming and filling state applies as well to the heavens as to the earth. Also it can be seen in this that there was another two part process involving first, the creation from nothing of the materials. The matter, the elements of the earth are first created in their proper proportion and then they are molded into the proper arraignment The work of creating them out of nothing was different from the work of molding them into shape.

Separate and arrange

Hand in hand with the forming and filling or after the forming and filling he would separate and arrange things. This is drawn from the word "divide or divided" as it is used throughout the chapter. He makes things and then placed them in their intended places in their proper proportions. Again a two part process can be seen. Creating the proper amounts and proportions of things was different from their locations or arrangement. This was the orderly and logical plan which he followed. The house was set up in this fashion.

The key of Evaluation or Stages

The key of evaluation comes from how we see that God checked his work as he went along. This comes from the phrase," and God saw that it was good". In this we can see that an aspect of creation was completed. Thus this phrase separates the creation up into different stages. Each evaluation represents completed aspect or stage of creation. The stages separate or break creation down into groups of related things. These groups are arranged according to how they relate to the forming and filling of the earth. For example the creation of light was a group or a stage unto itself which fulfilled the forming aspect of the light, whereas the animals were a group, fulfilling part of the filling of the earth with creatures. It should be noted that the stages are distinctive. That is separate from the others. Water creatures are separate from land creatures. The earth is separate from the heavens and so on.

The key of the Name

God named five things, Day, Night, Heaven, Seas and Earth. This is significant to what it all means to his overall intent and purpose. These are the things which he went about to bring into being, the day cycle and the three realms of sky, land and sea. This is the core aspects of our paradigm. We live a day to day life and we inhabit the Land Sea and air. We take them for granted. Other worlds do not have all these things as the earth does. The other planets do not appear to be completed as the earth was. It seems that they were formed in the raw material stage but never brought to a filled stage as the earth was. It did not serve the Lords purpose to fill them.

The key of the Earth

The bringing of the earth in to existence is a key to understanding the plan of creation, Both toward the other planets and toward the stars and toward the process of creation. We see the earth is formed first. The formation aspect of the earth was completed before any work was done in outer space. At the end of the fourth day the completed ecosphere of the earth stood alone. This reflects upon the principle of first things. Whenever something first appears in the scripture it usually affects all the rest of the occurrence of that thing. The first occurrence is the most important. The earth was first. The earth may not be at the center of the solar system but it is still central in Gods scheme for the universe and man. Maybe it is this principle of first things that helps explain why the sphere of the earth is made before the rest of the universe. Because the earth is the central place where the drama of the fall and redemption will take place. As a planet the earth stands in stark contrast to all other heavenly bodies at least as far as the purpose of God is involved... The other heavenly bodies are ther point in the creation process. The formation process of the universe is finished at the end of the third day. Next, came the filling process. How nicely it all fits together once you have discovered the grand scheme, guiding principle and ultimate purpose of God in creation. The earth is the principle place of creation and is crucial to comprehending why God went about things in the order which he did.

Key of Cycles
The day cycle is made as a part of making the first day. Then the day cycle is followed unto the completion of creation on the sixth day. This is followed by the addition of the Sabbath day. This seventh day is a memorial to the creation event; the totality of this adding up to seven. The day cycle was laid down and followed. Also the week pattern or cycle was laid down and followed. This is a chosen increment of completion in the mind of God. Just as with the earth being formed first which invokes the principle of first things so also this principle of first things applies to the first increments of time which God used in creation. The day and the week therefore are key cycles which man is to follow having been instituted by God himself in the creation event.

A study of numbers in Genesis 1
There are several repeating patterns of numbers in chapter one, which show us that God went about things in an orderly and chosen way. As we see them repeated, we may even come to expect them again. A certain type of prediction can be found in these repeating patterns of numbers. For example, each stage has three major divisions. This might not be apparent at first glance. However as you see this pattern emerge you then begin to expect to find this in some fashion in all the stages then upon further investigation you can discover it in all the stages. These things can be seen in listing the repeated numbers and patterns. In addition to these patterns it can also be seen to have a dominate role in nature. For example you would expect to find repeated groups of threes in the order of things. Plants, animals and even heavenly bodies should be fit neatly into groups of threes. Finally, these numbers and patterns even reflect back upon God, his nature and the chosen order of things. Take for example again the number three. Why all the grouping of three? Because, God is a threefold entity. The numbers 2,3,6,7 and 12 each have significant meaning and usages in Genesis.

Two is the number of grand divisions
God followed two grand principles in going about his creative works.
The principle of forming and filling [From verse 2]
The principle of creating and then arranging [From the word "divide"]
Each day of creation has two major events
. Day 1: The creation of light - separation of light from darkness
 Day 2: Creation of atmosphere - separation of the waters
 Day 3: Creation of dry land - Creation of plants.
 Day 4: Creation of heavenly bodies - Arraignment of heavenly bodies.
 Day 5: Creation of water life - Creation of fowls.

Day 6: Creation of animals - Creation of man.

The two fold basic cycle
The day is a two part creation. Creation has a light mode and a dark mode
. God often gets a double usage out of things. For example God created three realms, land, sea and air and each one of these has a light and a dark mode. Another example of this might be the two testaments of the Bible. In its completion, the scriptures are divided into these two grand groups. There are two genders, male and female. In verse, one the creation is divided into two major divisions The Heaven and the Earth.
Creation week can be divided into two major groups.
Three days of forming -days 1-3 Three days of filling - days 3-6 Other divisions of two might also be found in the scriptures. As in the carnal verses the spiritual or the two paths, the broad and narrow ways, and the physical body and the spiritual body.

The numbers three and seven
There are seven stages of creation. Each of these stages has a threefold division.
Stage 1- Light has three properties - Particle, wave and ?
Stage 2- The Earth has three realms - land, sea and air
Stage 3- Plants have three types - grasses, herbs and trees.
Stage 4- Heavenly bodies - planets, moons and stars.
Stage 5- Sea and air creatures - whales, other sea creatures, fowl.
Stage 6- Land creatures - cattle, beasts and creeping things.
Stage 7- Man, a threefold being - body, soul, spirit.
Seven in the number of completion, finality and perfection

In the broadest sense all thing fall into one of the seven stages. There are seven different broad groupings of things. For example light is a stage separate from the earth stage and the earth stage is separate from the plant stage and so on. The stages represent separate groupings. God made like things together and unlike things separate from each other. This is sound, practical and logical. There are seven completely different groups of things Three is the number of the "kind" or the number for types of things that are alike. Each separate stage of creation has three major categories. The groups of three are closely related as opposed to the stages, which represent the separation of things into distinctive groups.

Some implications of the numbers seven and three

One implication of this is that there are only seven major categories of things in the universe. This is because each individual stage represents a separate group of things that are unlike the others; one major category corresponding with each stage of creation. Each of these seven break down into groups of three related things. This makes for twenty-one major types of things. Three for each separate stage. As we take the world apart and put it back together this is what we should find. Another implication of this number is that creatures and other thing should be arranged into categories of three. If three is the number of "types" then as you examine nature one should find repeated groups of threes. As God, created things in major groups of three then it follow that these major groups will be broken down into small groups of three. This is a predictive implication of the number three in Genesis one. The patterning should follow on down the line. The kinds and the species should be discovered to fall into repeated groups of three. It follows that this should be scientific proof of the inspiration of Genesis. If all of nature is found to be this way then nature is verifying the patterning laid down in Genesis. This predictive patterning is a scientific principle, a law of nature revealed in the scriptures.

The tree of nature and life should look like the following

Stage - X First three - XXX Further divisions of three – XXX XXX XXX XXX XXX XXX XXX XXX XXX XXX XXX XXX Everything should fit nicely into one of the seven stages and then into a group of three. There is a word of caution in this idea of grouping. We know that God made the major grouping of three that are told to us in Genesis 1. However logically I must admit that just because God made the first groups of three it does not have to be that the rest are in groups of three. The logic is not of the type that it has to follow that everything else will be found to be that way. For example, you might find some combination of threes and sevens. But if the groups of three are found this is strong evidence for patterning laid down in Genesis. The other variation of what you might find then is that everything should relate back to its original group of three and then under this there could be much variation and diversification. Under this view the world would look like this: Stage- X Original three - XXX Diversification - XXXXXXXXXXXXXXXX Now as far as evolution goes there should be now patterning found in nature at all. All that you could expect from evolution is randomness and chaos. Because this is, all you could expect to find from an uncontrolled and accidentally process.

The number six is the number of man

This can be seen a couple of ways in Genesis, first by the total of the number of realm and modes of the earth. The earth was created with three realms, sea, land and air and two modes, light and dark. Together this makes six. The earth operates in a total of six modes. God made creatures to inhabit each of these. For example, there are creatures that operate on the land during the daytime and then others that come out at night. The same is done for the sea and the air. This is a total of six, secondly by the total number of days of creation. The work of creation was done in six days. On the seventh day God did not work. The earth and everything in it was made is six days. Finally, Man was created on the sixth day. His number is then related to the place that he lives, the earth that is; the number of modes of operation that there are in the earth and to the number of the day upon which he was created, the sixth day that is. This is a total of three ways that man is related to the number six. When taking into consideration the total number of modes or realms of the earth another interesting thought appears. This has to do with the number seven. Now as we have seen the earth has six realms; then if you add in the spiritual realm there are a total of seven. This fits perfectly with the idea of the number seven being the number of finality and completion. In the universe, there are a total of seven realms. Six are here upon the earth and the seventh in the spiritual world. By the way man's number being six adds support to the position of literal days of creation week. If man was not literally made on the sixth day then the number six is meaningless in regards to man's number. There may also be some relation to this found in the number of man as stated in Revelation 13:18 when there it speaks of the number of the name of the beast, 666, being also the number of a man.

The number twelve

Finally, the number twelve can be seen to relate to Genesis to some other things in nature and scripture. In creation week, there were 12 major action of creation two on each of the first six day. The year has 12 months. There are twelve cycles of the moon in one year and may in some way relate to creation. There are 12 notes in the musical scale, twelve tribes of Israel, and there were 12 disciples. I wonder if there is any relation to this and the way we buy our eggs today!

The age of the earth
The age or the earth issue is a grand side issue which goes right along with a person's views on creation. The split between the biblical and the scientific view has this issue as one of the grand differences between the two. The other main issue being the controversy about the sun being at the center of the solar system or not as brought to the forefront by Galileo.

It is too bad that these were not issue in the day of Jesus so that we could have received some direct comment from him on these matters. Nevertheless I believe that there may be a passage in the gospel of John which could shed some light on this age of the earth issue and offer some support to the position of a recent creation. In John chapter five verses one to nine is the story of Jesus where he healed the impotent man which lay by the pool of Bethesda. The possible inference to the age of the earth which I would make lies in this story.

To get the idea which I believe is imbedded in this text a person needs first to understand a few things. The first thing is that this miracle or healing the impotent man was a special sign given to the Jews that showed them Jesus was there Messiah. The second thing is that this impotent man was also a picture and a portrait of the helplessness of all mankind. The third thing it that every aspect of this man's life parallels the relationship of God with mankind since the fall of Adam.

Please let me lay out the case of this spiritual parallel to the reader because it is within this context that the inference to the age of the earth is found.

First the cause of this man plight was his sin. This is found in verse fourteen. *"Behold thou art made whole: sin no more."* The spiritual parallel is that sin is the cause of why he lay there. This sin of his reflects back to the sin of Adam and all our sin.

The second is that this man had no helper. Verse seven," *Sir I have no man when the water is troubled, to put me into the pool"*. In this is portrayed the helplessness of mankind to heal themselves of their sin problem. This man lay there among all the rest of the sick and sinful which represents the rest of mankind. Just as mankind was without a helper who could save them until Jesus came along. No one else was capable of the task.

The third is that Jesus was a person who could help him. Jesus was the man that he did not have. Verse eight" *Jesus saith unto him, arise take up thy bed and walk. And immediately the man was made whole and took up his bed and walked"* This portrays the salvation that is found in Christ through his death and resurrection

Now having seen how the life story of this man parallels and reflects the relationship of mankind since Adam up until Christ came let us notice what the scripture say about how long this man had laid there. Verse five says" *And a certain man was there, which had an infirmity thirty and eight years."* Now these thirty eight years in my view represents and reflects upon the time between Adam and Christ. We have to understand that in Gods economy every aspect of the encounter between Christ and this man has meaning and significance. So we must also include this thirty eight years. It was no coincidence that this man lay there that long or that Jesus just happened to pick him out of the crowd as an example. No he came there that day with that specific man in mind, healed him only and went his way because he was the portrait he was looking for.

Now having said all this to make the point that this thirty eight years represent in some way the time between Adam and Christ can we not also apply this to the age of the earth issue? Sure a believer in the inspired word of God should be able to. Since Adam was the first man who was made on the sixth day of the first week and the time between Adam and Christ is the same in some regard to the time between this impotent man and Christ, this same time also represent the time between creation and Christ.

What are we to make of this as regards to creation? Now we know that it could not have been thirty eight years between creation and Christ. After all this man was that old. Also we know as we extrapolate backward that it could not be three hundred eight years because Jewish history and the history of nations go back farther than this. But when we come to the figure three thousand eight hundred years now we may be getting somewhere when it come to the time frame between Adam and Christ or creation and Christ. This period could be supported by biblical chronology and the history of nations. Many notable people have made estimate of the time frame from creation to the present, bishop Usshers famous calculations for example. But in support of this idea that this thirty eight years which this man lay there represents a type of the figure of the time between Adam and Christ I would refer to the Jewish chronologer Rabbi Jose Ben Halafta. The time spans which he proposed from Adam to Christ was 1792 years which is very close in line with the calculation is in this impotent man scenario. And this is what I propose it also to be. That the time between Adam and Christ rounded off to the hundreds was thirty eight hundred years. Then of course we can figure from there up to the present time. All this makes the earth much younger that the modern scientific method would propose. But so be it. We may have a story from the very words of Jesus himself to support the young earth claim.

If you asked and Jew or most Jews living at the time how long the earth had existed and they gave you a general figure they would have said 3800 year. Also I can imagine how any Jew who understood the earth to be about 3800 years old and happened to be studying John chapter five could have made the same connection between this impotent man and the age of the earth as I have. So there you have my shot in the dark at the age of the earth May the Lord separate the wheat from the chaff!

The Rate issue
Move that Bus!
There is a popular show called Extreme Home Makes over, where they go around the nation and build new houses for people in need. These people are sent off on a vacation somewhere for one week and when they return there is a new house where there old one stood just a week ago. When they get the people back there the bus always sits between them and their new house and the host yells "move that bus" and there brand new house is revealed to them. It is amazing what can be done in a short time when you have enough planning and resources. Normally it takes six months to a year to build the average house. In addition, as you look around in the past sometimes many years are spent building some grand monument or cathedral. Now I can imagine as time goes by and say this show is forgotten about or all the information about how it was built was lost that if some future home owner was to be told that his house was built in a single week that he might not believe it, because normally houses are not built this fast. This brings up the "rate" issue. This is one of the hurdles that a person must face if they are going to accept the Genesis account of creation. God did not do an extreme make over but an extreme make-up of the universe. Just as a house can be built in, a week if you have enough resource so the universe could also given the power of God. The issue is, are the resource available, did and does God have such power. Apparently he does! Now I am amazed at all, they do to build that house in a week. Removing the old house, laying the concrete, getting it to dry so fast, framing, plumbing, electrical, heating and air, drywall, floors, kitchen, appliances, furniture, grass, drive way, landscape. Yes they do it all and even have time to film the show, all the planning of people and things, not getting in each other's way or holding each other up in along the way. Can we not expect that God could do the same? For many people it seems not. The question is

not if it can be done but are the resources available. Even many believers in him somehow are duped into accepting many of the tenets of the big bang cosmology not realizing that for one this is taking away from the glory of God and for another it violates what Genesis says. All of a sudden, we cannot decide what Genesis has said-pity. Why, we cannot even decide what the word "day" means, even though it is used hundreds of times in other places in the Old Testament! If we cannot decide on such a common word then what confidence can we have in the rest of the words of scripture? Really, why it is that we no longer agree with what Genesis says is because we want to appear educated or we want not to disagree with the so-called facts of science. This is an example of how we go about to find some place where we can bent or change what the scripture says and still look as if we are good old orthodox boys when we get done. This rate issue has a long history in the debate between science and the biblical chronology. There is the uniformitarian verses the catastrophic views of earth history; then there is the distance to the stars and age of the universe. All this depends on how fast or slow it was done. Some propose that all the layers of the earth were laid down over eons of time, that the stars that are so far away they must have taken so much time to get there. This all could have just have well taken place in a short time, but again it just depends on the rate and where they were in the first place. The present rate may not be the rate that they moved at some time in the past, the same goes for the layers of the earth. God has revealed to us the rate at which he worked. As incredible, as it seems he did it all in six literal twenty-four hour days. Why he took longer to build the earth that he did all the stars. The question is can mankind believe that he could work as such a rate or not? Part of the amazement of Extreme Home Make over is that all is done in a single week and part of the amazement of God is that he built our house it the same amount of time. This author for one refuses to take that away from him.

The cancer Illustration
A lesson on objectivity

Trying to get someone to believe in either evolution or creationism is like trying to convince people that cigarettes cause cancer. Now making a clear case that smoking cause's cancer for some is clear. The statistics show that many more people get cancer and other illnesses that those who do not smoke. At least the surgeon general thought so. However there are those individuals who have smoked many years and never developed cancer. It takes a long time for cancer to develop. {This might parallel the long ages that it take for evolution to accrue,} The way smoking effect the genes is not well understood. Making the direct link has been difficult. {Again the same could be said of evolution} Then to make matters even worse there are those who have never smoked and yet have had cancer. So there may be other factors involved. Smoking is not the only thing that can cause cancer; exposure to gasoline for example. So how is a person to decide? Or more to the point what is it that cause a person to decide one way or another over an issue where the facts {so-called} can be stacked up one way or another, or where there may be many varying circumstance and even far reaching implications? The answer is found in human choice. The answer is found more along the lines of what a person wants to believe than it is according to the facts that are involved in the case. We believe the one that we most want to be true. From this we can see that it is a heart issue. As much as we would like to think that we are all so objective and we see the world for what it really is. We fool ourselves by stating this or that fact in support of our case but really it is our heart and not our head that had made the decision. I notice that it is the people who smoke that are saying the things like. Uncle so and so smoked for fifty years and he never got cancer. But they never think that they don't have uncle whoever's genes. Or of the fact that just because he beat the odds doesn't mean that they are going to. The same thing might be said of getting a sexual disease. Or how about trying to convince people that their chances of hitting it big at the lottery are just not worth the effort or expense. Such is the case with evolution and creationism. A person I believe chooses which one he wants to be true just as much as he or she is convinced one way of the other by the facts. I wonder how much this want of the one or the other has to do with our decision? If we could place a percentage on it what would it be? Now for sure some people are more objective that others so there might be a range here. But what would it be? Fifty percent fact and fifty percent want; Sixty forty, or some other mixture of the two? What about if one or the other were paramount? I think we can rule out the being one hundred percent fact and zero percent want scenario. Our hearts are always going to have some part in what we decide. But I am not so sure that we can rule out the zero percent fact and one hundred percent want scenario. What I mean is that the want factor here becomes the all deciding factor. No matter how you line up the facts. No matter what the odds are, no matter what the facts are. No matter how good the arguments are. No matter how solid, No matter how sure. A person can still chose to believe something else; especially when the stakes are high and the higher the stakes the more the tendency to choose according to the want factor. In all the debating and all the arguing over this issue the one thing that is never brought up is how much a person wants the one, or the other to be true. Do we want God in our lives or not and how much do we want God in our lives or not is just as much a question that we should be asking ourselves as are any of the "facts". And in the final analysis this may be the only question that needs to be explored. We must become willing to cross the great divide of our wants, hopes and expectations before we can come to an honest and objective decision.

Pro 20:14 *"It is naught, it is naught, saith the buyer: but when he is gone his way, then he boasteth. Pro 17:8 A gift is as a precious stone in the eyes of him that hath it: whithersoever it turneth, it prospereth. Pro 11:1 A false balance is abomination to the LORD: but a just weight is his delight. Pro 16:2 All the ways of a man are clean in his own eyes; but the LORD weigheth the spirits.Pro 21:2 Every way of a man is right in his own eyes: but the LORD pondereth the hearts."*

David About the Author David Glaze holds a B.S. from Arlington Baptist College, 1984. He has preached, read and studied related books and materials on the creation and evolution debate for many years. He lives in Arlington Texas with his wife, Sherry. They have three sons, Zachary, Nathaniel and Caleb.

www.ingramcontent.com/pod-product-compliance
Lightning Source LLC
Chambersburg PA
CBHW030811180526
45163CB00003B/1237